T0326487

# Digital Informatics and Isotopic Biology

Self-organization and isotopically diverse systems
in physics, biology and technology

# Digital Informatics and Isotopic Biology

Self-organization and isotopically diverse systems
in physics, biology and technology

**Alexander Berezin**

*Professor Emeritus, McMaster University, Hamilton, Ontario, Canada*

**IOP** Publishing, Bristol, UK

ISBN    978-0-7503-1293-6 (ebook)
ISBN    978-0-7503-1294-3 (print)
ISBN    978-0-7503-1295-0 (mobi)

DOI    10.1088/978-0-7503-1293-6

Version: 20160801

IOP Expanding Physics
ISSN 2053-2563 (online)
ISSN 2054-7315 (print)

British Library Cataloguing-in-Publication Data: A catalogue record for this book is available from the British Library.

Published by IOP Publishing, wholly owned by The Institute of Physics, London

IOP Publishing, Temple Circus, Temple Way, Bristol, BS1 6HG, UK

US Office: IOP Publishing, Inc., 190 North Independence Mall West, Suite 601, Philadelphia, PA 19106, USA

# Contents

# Preface

This book was written by a theoretical physicist who was active in several areas of physical and mathematical sciences and suggested a few ideas at the interface of quantum physics, biology and informatics. The book is highly interdisciplinary in its scope, approaches and inferences. One of the primary imports of this book is the discussion of numerous aspects of the isotopic diversity of chemical elements under the umbrella term *isotopicity*. In 1984 this author proposed the hypothesis of an *isotopic biology*—an alternative genetic code based on the combination of stable isotopes in bio-informational structures. While up till now this hypothesis has remained unconfirmed (and no attempts to study it experimentally are known to this author), its potential ramifications could extend to several areas of the biological, bio-medical and ecological sciences, as well as to new informational systems (digital informatics and quantum computing). Much of the narrative is accompanied by some philosophical reflections on issues of fundamental importance for the discussion.

Special emphasis is placed on the informational ('digital') aspects of isotopic randomness and isotopic self-organization. Everything that we can see and touch—including ourselves—is made of atoms and about 2/3 of all chemical elements have two or more stable isotopes. There are 254 known stable isotopes and 80 elements in the periodic table that have at least one stable isotope each. Twenty-six elements only have one stable isotope. These elements are called monoisotopic. They are, as it were, a minority among all the stable elements. The number of radioactive isotopes—occurring naturally and created artificially—is much higher, with there being some 2400, or so.

The above facts are, of course, generally well known. However, from the start there appears to be a somewhat curious paradox in the realm of science and technology. On the one hand, the existence of isotopes and their numerous applications is common knowledge, part of the intensive and versatile research activity that has gone on for over a century, and as such, the science of isotopes forms a significant and clearly visible corner of physics, chemistry and material science, to name just a few major outlets. There are several major industries and technologies that are critically dependent on isotopes, including socially and economically important areas such as nuclear science and engineering, medicine (where isotopes are widely used in treatment and diagnostics), isotopic geology (e.g. the dating of minerals and sediments), pollution and radiation monitoring, forensics, and so on. These technologies are the basis for huge businesses that employ many thousands of professionals with various skills and training, and, cumulatively, they account for a substantial share of the modern economy. Likewise, it goes without saying that all aspects of isotopic science and technology have produced a massive and ever-growing body of literature, shelves of books and PhD dissertations, with innumerable research careers being dedicated to the study of isotopes and their applications.

However, on the other hand, there seems to be little appreciation for isotopic diversity as a singular phenomenon of the natural world, a phenomenon of its own kind, with a distinct and unique signature and a broad range of manifestations in the material world. Such a unifying (one might call it 'holistic') view of isotopic diversity and randomness is the *leitmotiv* of this book.

A large fraction of my scientific work during my 30-year teaching career at McMaster University in Hamilton, Canada (1980–2010) was devoted to various aspects of the diversity of stable isotopes—an area which I, provisionally perhaps, call *isotopicity*. If anyone asks me whether I really am the original creator and user of this term—that I cannot assert with full confidence. In an ever-increasing ocean of information, almost no one can lay a claim of this kind with full confidence, especially in the area of terminology. Suffice it to say that, to the best of my knowledge, no one used the term isotopicity before me. I have published some 50-plus papers on isotopes and the present book is, to some degree, a compilation of and a follow-up to this work. Not everything in this book is an easy read, but almost all of it is thought-provoking. And never mind that certain things may be repeated more than once. Unlike works of fiction, which are read for entertainment, non-fiction books are read (mostly) for their ideas and intellectual stimulation.

Even occasional fierce criticism and outright rejections of my scholarly offerings have, paradoxically, often stimulated me to sharpen my metaphysical argumentation and filled me with a zest to carry on further and into more extended 'scholarly fights'. Some of them were successful, some maybe not, but all were, in my view, well worth fighting. And there is the standing possibility that perhaps some of my ideas on isotopicity and the other issues mentioned in this book will be picked up and extended 50, or 100, or 1000 years hence (or, perhaps in some parallel universe with multidimensional time?), which gives me another (metaphysical, so to speak) incentive to scribble these notes.

To restate, this book emphasizes the concept of isotopicity as a singular aspect of nature with numerous physical, technological and biological applications and implications. I do not raise special claims with regard to the originality of the ideas discussed in this book. On the contrary, I see them to a large degree as restatements or sharpenings of previously circulating notions and ideas. For example, the idea of isotopic biology is, in a nutshell, a hypothesis that subtle information-carrying genetic messages can be 'overwritten' in isotopic permutations in the chains of genetic codes (DNA and other similar polymer molecules). This idea was expressed in several of my papers from 1984–87 and to the best of my knowledge it may indeed be an original one.

While I do not stress with absolute certainty that no one before me spelled out similar ideas, searching through the literature and the web has not uncovered any prior discussions to this effect (about the bio-informational role of isotopic diversity), there are only a few follow-up discussions with reference to my original publications. But, again, I do not claim that my searches have been comprehensive, it may well be that some comments or suggestions of this kind were spelled out earlier. Should such comments ever come to my attention, I will not be embarrassed in any way or form, since I lay out no 'priority claims' of any kind. On the contrary, if any such previously stated ideas concerning isotopes become known to me, I will take them as additional confirmation that I was (am) on the right track regarding the ideas of isotopicity as presented in this book.

And in the spirit of *A Mathematician's Apology* by Godfrey H Hardy (1940), I feel I should also (sort of) make my 'apology' for the fact that many (well, it may appear, too many) references in this book are to my own papers. Whether some readers may take this as a sign of vanity or self-promotion is not for me to predict, I only hope that they will not. I have never sought public acclaim and have never been presented with any titular laurels (awards, honorary titles, etc) for my work, but the fact that in some sections of my book almost all the references are to my own papers is due to my lack of knowledge of similar work by other people—in cases where I was aware of such work, I certainly included references to such publications.

This books talks a lot about digital strings, prime numbers, infinite sets and all other mathematical wonders. In this way, I see myself as a follower of the Pythagorean–Platonic tradition that considers mathematics (numbers, forms, etc) as the true foundation of the Universe. And our present ongoing (and expanding) immersion into the digital world amply confirms this. If I had to define the quintessence of the present stage of technological civilization, I would suggest *digital Pythagoreanism* as a term that reflects the essence and zeitgeist of our age.

*Alexander Berezin,* June 2016

# Foreword

This book was written by a theoretical physicist who was active in several areas of physical and mathematical sciences and suggested a few ideas at the interface of quantum physics, biology and informatics. The book is highly interdisciplinary in its scope, approaches and inferences. One of the primary imports of this book is the discussion of numerous aspects of the isotopic diversity of chemical elements under the umbrella term *isotopicity*. In 1984 this author proposed the hypothesis of an *isotopic biology*—an alternative genetic code based on the combination of stable isotopes in bio-informational structures. While up till now this hypothesis has remained unconfirmed (and no attempts to study it experimentally are known to this author), its potential ramifications could extend to several areas of the biological, bio-medical and ecological sciences, as well as to new informational systems (digital informatics and quantum computing). Much of the narrative is accompanied by some philosophical reflections on issues of fundamental importance for the discussion.

This author makes no excuses for such philosophical (and often 'metaphysical') reflections and deflection, as he is profoundly aware that in a research community and, in particular, a physical community, there are three dominant attitudes with respect to such philosophical detours. One is an utter disdain and distaste for anything that smells of philosophy or, God forbid, metaphysics: 'we are physicists, we only deal with and care about what we observe experimentally and can measure with our apparatus'. The second attitude, which I would call the middle one, is somewhat neutral—basically, showing little interest in things philosophical; such remarks can be ignored or skipped. And, finally, the third attitude is one of keen and profound interest, showing a sense of awe and excitement that the present author has often observed during his numerous talks at conferences and seminar presentations. Fortunately, this last community (to which this author himself belongs) is pretty numerous and this is the community to which this book is primarily addressed. Another redeeming point here is that there is often a strong interest in these ideas among various lay people who do not formally belong to any specific sciences and, again, from this author's observations, such a cohort seems to be quite numerous and diverse.

The book begins with a review of the duality of energy and information as prime physical categories of the world. The conundrum of chaos versus order is reviewed with some examples from the physics of self-organization, spontaneous symmetry breaking and structural phase transitions in the systems of interaction particles. These discussions are largely based on the work of this author, as published in numerous papers. Spontaneous symmetry breaking and the spontaneous emergence of order in multiparticle systems is one of the primary facets of complexity theory, with numerous ramifications for such areas as the theory of evolution, the origins of life and the nature of consciousness. The present author adds his modest contribution to these ideas by tackling hopping conductivity and quantum tunneling in crystals with impurity centers and the quantum effect of Anderson localization in disordered crystal lattices.

A biographical digression will explain how this author came from his studies on hopping conductivity and crystal symmetries to his inferences and hypotheses on isotopicity and isotopic biology. An overview of the nature of discrete information carriers (digital strings) and normal numbers (real numbers that contain any combination of digits) is given, and this discussion offers some philosophical glimpses of Georg Cantor's theory of infinite sets, while also featuring the vivid metaphor of the *Library of Babel* from the well-known short story of that name by Jorge Luis Borges.

The chapter on isotopicity and isotopic randomness points to a somewhat unusual situation in the research community regarding understanding the role of the diversity of stable isotopes in a variety of physical and biological effects. Most people with even an elementary knowledge of science are almost certainly familiar with the word 'isotope'. This word almost always invokes the notion of radio-activity, nuclear energy, nuclear medicine and the like. These areas are based primarily on *radioactive* isotopes—atoms whose nuclei are unstable. At the same time it is somewhat less appreciated that most chemical elements exist in the form of two or more *stable* isotopes. For example, the prime chemical element of life—carbon—is a mixture of two kinds of atoms called carbon-12 and carbon-13. Nuclear technology, nuclear medicine and numerous other areas of modern life are critically dependent on isotopes. Since the main elements of biological life are carbon, oxygen, nitrogen and hydrogen, we (our bodies) are all mixtures of stable isotopes (with a trace amount of radioactive isotopes, like carbon-14 and potassium-40). This book is largely focused on *stable isotopes* and their numerous implications and applications.

To give but one example of how isotopicity may affect the biological realm, consider a long polymeric molecule of DNA—something which is common to all of us and to almost all known life forms. It is a long polymeric chain of several chemical elements, mostly carbon, oxygen, nitrogen and hydrogen. (It is estimated that if all the DNA molecules in our body could be stretched into a single string, its length would be some 1.2 billion kilometers—four return trips to the Sun.) The intricate order and the arrangement of DNA strings encodes all the vital operating instructions for all aspects of the biological dynamics of living systems. This marvelous informational system—DNA's genetic code—is the centerpiece of all Earth's biology, and since its discovery in 1950s we have accumulated a vast volume of knowledge about its structure and functioning. Far from just being a subject of passive academic knowledge, DNA code is now a foundation of various facets of genetic engineering, medical and forensic applications, and other areas of emerging genetic technology.

How does the isotopic diversity of elements affect all that? As a couple dozen letters of the alphabet can be used to 'code' any text through the proper ordering of letters and spaces, genetic DNA code has all the information stored through the proper ordering of the atoms and molecular units (codons) and their combinations. However, the key atoms of DNA have different stable isotopes. Hydrogen, carbon and nitrogen each have two—H and D, $^{12}$C and $^{13}$C, and $^{14}$N and $^{15}$N, respectively—while oxygen has three, $^{16}$O, $^{17}$O and $^{18}$O. The question arises of whether a specific order of isotopes can carry additional (or alternative) information that is overwritten above the 'regular' message of DNA, with the latter being stored in the spatial arrangement of DNA atoms. An analogy for this are the so-called 'subtle messages' that are sometimes claimed to have been overwritten on some pieces of music. In other words, isotopic diversity can be seen as an *additional level of information capacity* over and above the regular chemical diversity of elements. Likewise, the purposeful ordering of isotopes in crystalline structures can be used for alternative information storage systems, as was proposed by the present author in a series of papers (and, of course, this is discussed in detail in this book).

This book discusses the numerous implications and applications of isotopic diversity for physics, biology, geology, material science, informatics and information technology, including such novel areas as nanotechnology and quantum computing. It also ventures into such controversial areas as the idea of water memory (can water store memory in isotopic patterns?), the formation of 'isotopic neural networks' in crystals (does this have anything to do with the alleged 'healing effects' of quartz crystals?) and the potential use of isotopic combinations for alternative methods of energy generation (there is an enormous quantity of hidden energy stored in isotopically mixed structures). There is also a discussion of a possible role for isotopic randomness in the mechanism of consciousness, as was proposed by the present author in a series of papers published in a variety of journals, such as the *Journal of Theoretical Biology*.

The key point here is that recent ideas on the quantum nature of consciousness (see Roger Penrose and others) afford the central role to the process of the quantum reduction of the wave function (quantum collapse) of the neural structures that is induced by gravitational fields. Since isotopes have different masses, their gravitational potentials are slightly different, and that makes the processes of the said quantum collapse sensitive to the isotopic environment. The details of this are outlined in the book in a qualitative (non-technical) manner.

As a general remark, the author suggests that the prime philosophical quest behind the concept of isotopic biology can be formulated in the following way. If Nature is smart enough to use the diversity of chemical elements for biology (almost all the elements in the periodic table have some biological functions, including microelements), then it may look somewhat odd that Nature would omit to use such a mighty additional informationally rich resource as the diversity of stable isotopes for the structuring and functions of biological systems at all levels of evolution and complexity. The likely 'answer' to such a 'puzzle' is that, yes, Nature most likely uses it (isotopic diversity), but we have so far failed to detect this and have even (largely) failed to look at it, even at the level of a hypothesis, not to mention through targeted experimentation. One of the prime aims of this book is to draw the attention of the world's research community to this incipient research area of stable isotopicity and isotopic engineering—a direction that (with some luck) may turn out to be a newly found gold mine for physics, biology, biomedicine, material science, cognitive sciences and informational technology in a broader sense.

The book is extensively interdisciplinary and holistic in its nature and intent (in the sense of the 'holomovement' of David Bohm). My intention was to make it profoundly thought-provoking and stimulating, especially for young minds who are still searching for their own path and place in science and research. My previous experience in presenting these ideas has quite consistently demonstrated that there are, generally, two almost opposite attitudes when facing these ideas. On one side, there is the open-eyed curiosity and amazement of the 'what if?' type of reaction, which is often mixed with puzzlement and awe; this is the 'tell us more about that' perspective that comes largely (but not exclusively) from a younger audience. On the other side, there is the disbelief and skepticism of the 'it cannot be so, because it cannot' mindset, which is mostly (but, again, not exclusively) voiced by so-called 'mainstream' and often senior scientists (the 'we know better' syndrome), some of whom are often immediately dismissive of anything that (in their opinion) 'does not fit' and trespasses against the boundaries of the established consensus of science (again, the way they see it). I talk more about this in the concluding chapter of this book.

The concluding chapter addresses some historical and social issues that are relevant to the themes of the book. In particular, it proposes the analysis of the paradoxical (in the view of this author) situation where in spite of it being one of the prime facets of the material world, stable isotopicity and the effects of isotopic randomness receive very little attention from the research and technological community. In the view of the author, this neglect and the lack of serious research activity can be attributed to a large degree to the present structure and *modus operandi* of the modern research system, which is based on 'grantsmanship' and 'expert peer review'. The latter is largely unsupportive of and often directly hostile towards novel and risky ideas. The net result is the often seen group mentality and the presence of all sorts of conformist pressures on researchers, especially those who venture to deflect from the dominant consensus ('do and think as we do'). All this translates into the encouragement of incremental research along well-established research lines and the suppression and dismissal of new ideas and innovative offerings.

# Acknowledgements

Most prefaces to books end with 'acknowledgements'. My prime acknowledgement is to my late wife Irene (1943–2005), with whom I raised two wonderful children and whose love and care for over 37 years of our marriage have given me the stability, energy and motivation for my diverse scholarly pursuits in physics, metaphysics, the arts, museums, travelling and many other things that we accomplished together. In addition, many other people, whether they were professional scholars or not, have shown interest in my work and given me encouraging words and useful comments that often helped me along. I remain thankful to all of them, but in order not to disfavor anyone (because it is impossible for me to recall all of them), I shall refrain from giving any names here.

# Author biography

## Alexander Berezin

 Alexander (Alex and/or Sasha) Berezin was born 26 April 1944 in Russia (USSR), during the last year of the Second World War. Until 1978 he lived in Leningrad (now Saint Petersburg) and in 1970 obtained a PhD in theoretical physics (quantum solid state physics) from Leningrad State University. Apart from physics he received an education in the History of Arts from the Hermitage Museum in Leningrad where his mother worked as a senior expert in arts and a curator of the French painting collection. From 1969 to 1974 he worked as a researcher (theoretical physicist) at the Ioffe Physical–Technical Institute of the Academy of Sciences of the USSR, Leningrad, and from 1974 to 1977 he was a Docent (Associate Professor) at the Department of Physics, Makarov Higher Naval Engineering Academy, Leningrad. In 1978 he emigrated with his family (wife and two children) to Canada. Between 1978 and 1980 he was a research associate at the Department of Physics, University of Alberta, Edmonton, Alberta. His last work (1980–2010) was as Professor, Department of Engineering Physics, McMaster University, Hamilton, Ontario, and he retired in 2010. His present status is Professor Emeritus and he lives in Toronto.

His areas of interest include: quantum physics, nanotechnology, physics of information (energy and information), electromagnetism, electrical and thermal properties of materials, physics of isotopes, extended ideas on isotopic diversity (isotopicity) in digital informatics, biology, biophysics and biomedicine and emerging technologies, non-equilibrium thermodynamics and physics of chaos, randomness and self-organization, randomness and creativity, complexity theory, phase transitions and catastrophe theory, quantum computing, non-local quantum effects and quantum entanglement, singular (delta) potentials in quantum mechanics, environment and health physics, sustainability, electrostatics and self-organization in Coulombic systems, pollution control systems (electrostatic precipitation), quantum physics of consciousness, physics of homeopathy (memory effect in solutions) and 'healing crystals', philosophy and foundations of mathematics (Platonism and Pythagorism, Cantor's 'alephs', number theory, prime numbers, tower exponents), fractals and Mandelbrot sets, cosmology, 'parallel universes' and inflationary cosmological models, virtual and simulated realities, transhumanism, singularity, infinity and universal informatics, *Library of Babel* by Jorge Luis Borges, the arts (in particular, visionary arts and surrealism), architecture and design.

He has published 160 peer-reviewed papers in major physical and engineering journals, 65 articles in conference proceedings and 260 other publications (abstracts of talks and seminars, magazine and newspaper articles, reports, etc).

**IOP** Publishing

Digital Informatics and Isotopic Biology
Self-organization and isotopically diverse systems in physics, biology and technology
**Alexander Berezin**

# Chapter 1

## Introduction

*In order to more fully understand this reality, we must take into account other dimensions of a broader reality.*

John Archibald Wheeler (1911–2008)

*There is a fifth dimension beyond that which is known to man. It is a dimension as vast as space and as timeless as infinity. It is the middle ground between light and shadow, between science and superstition, and it lies between the pit of man's fears and the summit of his knowledge. This is the dimension of imagination. It is an area which we call the Twilight Zone.*

Rod Serling (1924–1975)

The lives of great people are often surrounded by a variety of short stories and the smart words ascribed to them. Richard Feynman was once asked (or, perhaps, he asked the question himself), 'if, in some cataclysm, all of scientific knowledge were to be destroyed, and only one sentence passed on to the next generation of creatures, what statement would contain the most information in the fewest words?' His answer was 'I believe it is the atomic hypothesis that *all things are made of atoms— little particles that move around in perpetual motion, attracting each other when they are a little distance apart, but repelling upon being squeezed into one another*. In that one sentence, you will see, there is an enormous amount of information about the world, if just a little imagination and thinking are applied'.

This idea of atoms as fundamental and tiny building blocks from which everything is constructed was with us long before we invented any instrumentation to discover their existence. It is sufficient to recall the great ancient atomists (Leucippus, Democritus, Epicurus, Lucretius—and, most certainly, there were many more) who talked about the atomic world, sometimes with amazing insight and imagination. Skipping through centuries we come to a revival of atomistic ideas in such figures of pre-experimental atomism as Giordano Bruno, René Descartes and Robert Boyle, to name just a few.

Over the last two centuries, enormous progress has been made in all the natural sciences (physics, chemistry, biology, geology, cosmology, etc), and the mighty technological structure that humanity has built has put the notion of atoms, molecules and chemical compounds (many thousands of them, both natural and synthesized) into the category of trivia that is automatically known to everybody from kindergarten upwards.

However, as we often hear, the devil is in the detail. There are many subtleties and as we well know, confusion and misunderstanding exist about many things. Moreover, the confusions, misunderstandings and aberrations are often contagious and stubborn. Nonetheless, the spectrum is quite mixed here and we equally well know that disagreements, disputes and discourses regarding the key issues can often be healthy, constructive and insightful. In the view of this author, in the scientific community there is some noticeable degree of confusion, misunderstanding and underappreciation regarding the possible role and place of the diversity of stable isotopes in the physical and biological world. This book (and the author's prior work) aims to fill this gap. The author uses the umbrella term 'isotopicity' to describe the phenomenon of the existence of isotopes and the numerous effects coming from such a diversity (and isotopic randomness), and the discussion of the many implications and applications of isotopicity constitutes the *leitmotiv* of this book.

So, why isotopes and isotopic randomness? What can be said about them that has not already been said?

Most people with even an elementary knowledge of science are probably familiar with the word 'isotope'. In the public perception, this word almost always invokes radioactivity, nuclear energy, nuclear medicine and the like. These areas are based on radioactive isotopes, atoms whose nuclei are unstable. And because we all (to various degrees) are mesmerized by the words and terms (a Pavlovian conditional reflex), for too many people, if not for the majority, hearing the very word isotope(s) produces some reaction of distaste and even fear. For many people, isotopes, both consciously and subconsciously, trigger an instant association with radioactivity (which is generally perceived as being environmentally harmful), not to mention nuclear weapons, dirty bombs and similar less than pleasant things. Never mind that stable isotopes have nothing to do with any of the above—to the public (including even many scientists, who are supposed to know better) it is the 'i-word' and is best avoided. That is why when you mention working in the area of isotopes, it too often invokes an image of a 'mad scientist'—the reaction that I myself, perhaps, have faced a number of times ('could you not find anything better to do?').

The somewhat negative aura surrounding everything related to isotopes explains, at least partially, some other public misunderstandings. The focus on radioactive isotopes leads, by the law of unintended consequences, to the net result that the existence of stable isotopes for many chemical elements is far less appreciated by the public than the actual facts of science call for. For example, the prime chemical element of life—carbon—is a mixture of two kinds of atoms called carbon-12 and carbon-13. But, as was just said, because nuclear technology and nuclear medicine (both are major components of modern life) use radioactive isotopes, this somehow shifts public attention away from the proper appreciation of stable isotopes, which in this way take a back seat at best—despite the fact that the whole world we see around us (including ourselves) is made up almost exclusively of stable isotopes. This is perhaps the main reason why the notion of isotopicity as a single concept has received (at least so far) relatively little attention (if not a denial) from the science and technology community.

The author of this book has been working to fill this gap for some 25 years and what follows is largely a compilation of various offerings and ideas from scores of journal papers published over the years. The author believes that the explication of the notion of isotopicity—the isotopic diversity of chemical elements—is conceptually and paradigmatically beneficial and brings a new and a viable 'meme' into scientific discourse. Whether isotopicity as a unifying concept is going to take its rightful seat alongside other paradigms in the sense of Thomas Kuhn's *Paradigm Shift* (Kuhn 1962) will be seen in the coming years; I, as an author, do not make firm predictions on this matter.

The physical effects, technological options and ideas discussed in this book relate almost exclusively to the stable (non-radioactive) isotopes of chemical elements—the very fact that most (about 2/3) chemical elements have two or more stable isotopes opens the door to many interesting effects and applications, some of which are well known, but others of which can only be indicated tentatively at this stage. Focusing on stable isotopes leaves out such important areas as nuclear energy, nuclear medicine and nuclear weapons. All these areas are already well covered, as reflected by the vast existing literature (not to mention electronic resources) and, hence, there is no need for me to review these topics. My preference is to concentrate on less trivial aspects and ideas of isotopic diversity in nature and technology, as well as on some possible links of isotopicity to greater realms of ideas.

The text which follows makes the case for isotopicity (as a singular unified concept of its own) in a dual sense. On the one hand, I go through many physical effects and technological suggestions based on my expertise in quantum physics, condensed matter physics and material science. On the other hand, this concept of isotopicity, in the way I propose it, falls in the line of a metaphysical discourse. It may even have, perhaps, some flavor of a surrealist vision. My own vision of isotopes, and the physical world in general, invariably had (and has) a surrealist tilt and with all due humility I make no apology for bringing up such an inference.

That said, I would like to think that one of the impressions that this book may bring to the reader will be akin to the impression one might carry out after a tour of some large museum of modern art. Such a museum will normally exhibit a mixture of styles, qualities and ranges, from, say, realistic landscape painting through to abstractionism and surrealism of the Salvador Dali kind. Personally, I am a fan of Salvador Dali and I take a lot of inspiration from his art. There are mountains of books about surrealism and Salvador Dali in particular (e.g., Dali 1942, Cowles 1959, Ades 1982, Descharnes and Neret 2001, Salber, 2004). In fact, it is almost a common point for theoretical physicists to be inspired by (and often admire) surrealistic art, as represented by such artists as Salvador Dali, Yves Tanguy, Maurits Escher, Rene Magritte, Paul Delvaux, Giorgio De Chirico and many others. In some ways, my work in science, as well as the structure and presentation mode of this book, have some 'surrealistic spirit' in them. Isotopic randomness, the theory of chaos, random digital strings, isotopic self-organization and pattern formation—all of these intellectual goodies have surrealistic overtones (for those who want to see them there, of course).

Because of this book's 'exhibitionist' character, I have not been particularly meticulous about some repetitions of the same or similar thoughts and ideas creeping in here and there. Some of them have been taken from various (usually edited) papers of mine. As visual artists often go through repetitions, usually with variations, of the same or similar themes, so do I. My primary goal is to convey the general flavor of isotopicity and the many facets of isotopic diversity in nature and technology. For that matter, a somewhat chaotic style of presentation of the material in this book seems to me to be a good choice, especially because I, as a physicist, love chaos theory and made some—albeit quite modest—contributions to it, mostly in the 1990s.

This book discusses a broad range of physical effects related to isotopes and isotopic randomness. It also outlines numerous ideas that range from quantum solid state physics through to biology, informatics, environmental areas, bio-medics, and even some psychology and sociology. The ideas presented in this book involve various degrees of hypothesizing (or some may prefer the term 'speculation') for which I, as an author, make no apology. No science, and almost no intellectual development, can do well without some measure of hypotheses and speculations. Nowhere in my book do I make any proselyting or authoritative statements regarding isotopicity and the effects that may (or may not) be related to isotopic diversity, and in no way do I proclaim any final truths here. As one of the reviewers of my book pointed out, this book poses far more questions than it provides definite answers. This open-endedness with its thought-provoking style constitutes, in my humble opinion, the most attractive aspect of this book for potential readers.

Many of the issues discussed in this book have overlaps with one another and there may be textual repetitions and redundancies. Again, as an author and a long-time lecturer, I make no apology for this, recalling a well-known Russian saying, *povtorenie mat' uchenija* (repetition is the mother of learning). The way the book is structured is selective-reading friendly and different chapters and sections may well be read on their own or in any order. In fact, numerous sections of this book can be read as (almost) independent essays. That, in my view, justifies some degree of redundancy and repetitions in the text. In my own experience, non-fiction books are rather rarely read from cover to cover by either professionals or amateurs. This is understandable in view of there being so many books around (plus such vast web resources), which is one of the principal characteristics of present-day civilization, even though people have complained about the same thing for centuries.

Some of the effects discussed in this book have not yet been experimentally tested (or may not have been?) and, hence, much of the discussion remains at the level of hypotheses. The selection of effects and research offerings discussed to a certain degree reflects my personal preferences and interdisciplinary interests as an author, as well as my prior experience in the area of isotopicity and isotopic engineering. I must admit that in recent years, especially after my retirement in 2010, I have not followed experimental and theoretical developments in physics carefully and certain effects that I describe as 'hypothetical' here may, in fact, have already been discovered (or dismissed?) by other researchers in the area. Again, I make no apologies of any kind for that because the last thing I want is to do is to lay any claims to priority or to pretend to be a pioneer[1]. This book should not be seen as an attempt to give an exhaustive and up-to-date review of all the recent data on isotopes and their applications. In view of the enormous and exponentially growing array of information, providing an exhaustive review would not only be virtually unattainable, but it would also quickly become incomplete, with new data coming in almost daily. This is probably also the case for practically any area of an active research.

Thus, the actual (and in my view, reasonably attainable) target of this book is to focus attention on the rich variety of ideas related to the notion of isotopicity and other issues stemming from it. Some of these issues (like digital informatics, or some philosophical aspects of randomness) may have a varying degree of closeness to the prime issue of isotopic diversity, yet the overall primary goal of this book is to stimulate further thinking and the exploration of the issues discussed in it.

The list of references in this book, while it refers to numerous articles and books, may be somewhat shorter than reference lists in many other non-fiction books. I have seen non-fiction books in which about one quarter (if not more) of the book is taken up by comments and references. However, with the ever-increasing power of the internet that we all witness daily (exponential Moore's law), and with such powerful search engines as Google, any interested reader can easily go to the primary sources.

Likewise, some quotations and ideas were taken from a variety of web sources, including Wikipedia. In all these cases, I edited and performed reasonable validity checks, but formal referencing in some such cases is impractical and, for practical reasons, often unrealistic. All this reflects the changing practice of modern scientific writing, where numerous web sources provide an effective compensation for the incompleteness (or even a lack) of formal referencing. The archives of most scholarly journals are now available online from (almost) any point in the world, although some readers may still face problems of access (not everything is free in this world, at least not yet).

As an author, I fully realize (as anyone else should) that the world neither begins, nor ends with me. Sometimes (quite, rarely, though) books outlive their authors. There are millions upon millions of books in the world. As Wikipedia says, the Library of Congress, which is said to be the largest library in the world, holds some 140 million items and 33 million books on approximately 830 miles (1335 km) of bookshelves. And these are by no means all the books ever published in the world, as many have not made it to the Library of Congress (the total number of books published in the world is estimated to be between some 100 and 300 million and this number is currently growing at the rate of some 6000 new book titles per day)[2].

---

[1] Although I actually was one, in a manner of speaking—in my childhood in the Soviet Union in the 1950s all schoolchildren were 'pioneers', wearing mandatory red ties. You can still find such images on the web, look for the Soviet art of the 1930s–1950s.

[2] Source: https://en.wikipedia.org/wiki/Library_of_Congress

Some authors, even in the early days of book printing (say, in the 16th century), were already complaining of there being too many books. And yet, people keep writing and publishing. In the light of all this, I make no predictions of any kind about this particular book and allow myself no specific expectations. To quote Ludwig Wittgenstein (26 April 1889–29 April 1951), with whom I share a birthday (mine is 26 April 1944), 'if there ever will be a single reader of my book, I will eternally be satisfied'[3].

Another issue I would like to address in this introduction (and later in chapter 6) is the system of peer review. While it has both (alleged or real) merits and flaws, the modern peer review system has quite a noticeable impact on the dynamics of scientific practice in terms of both research funding and publication decisions. My long (over 40 years) personal experience with peer review has made me profoundly critical.

Over the years I have had numerous clashes with the peer review system in science, for both publication and research funding. The latter refers to the process of awarding research grants by funding agencies, like the Canadian NSERC—Natural Sciences and Engineering Research Council of Canada. I have published many (over 20) articles and letters that criticize peer review, which I personally see as being much more damaging than beneficial for science. For example, in my paper 'Discouragement of innovation by overcompetitive research funding' (Berezin 2001), I outlined the stifling and suppressive effect of peer review—the system that is inherently prone to numerous conflicts of interests, obscurantism, plain jealousy and (often almost paranoiac) hostility—all of which is covered by the traditional peer-review anonymity. Of course, and by all means, I am far from being alone in such a criticism—numerous scientists and authors have voiced similar concerns.

During my most active period of science writing (from 1967 to about 2010) I received many rejections for my papers from a variety of journals. However, by persistently resubmitting them to other (and sometimes the same) journals, I eventually succeeded in publishing almost everything I wrote. For the rejected papers, the comments from reviewers were invariably pointless, often sarcastic and hostile, and possessed almost no usefulness for me in any constructive way. Even those reviewers who appeared to be positive and recommended that I make changes (recommended a revision) were rarely really useful for me. In those cases, when I was making the suggested (in fact, required) revisions, I, for the most part had to 'soften' some of my most controversial statements. In other words, these reviewers forced me to make some compromises with my ideas, although (fortunately) I was still able to get around these difficulties and convey the essence of my messages undistorted. In summary, I can hardly recall even a single case when reviewer comments were really useful for me. Except, ironically, in giving me plenty of material with which to write papers (peer-reviewed, of course) about the detrimental qualities of the peer-review system (e.g., Berezin 1996, 1998a, 2001).

Certainly, as in almost any system, there may be some exceptions. There are people on all levels of the scientific ladder who are not constrained by tunnel vision or the

---

[3] Wittgenstein L 1921 *Tractatus Logico-Philosophicus*

'not in my backyard' mentality and can see outside the box. However, there are relatively few people with a global and interdisciplinary vision in the science community (in my experience, at least) and they may not be favored as peer reviewers that often. But, again, the effect of randomness may play some positive role here.

Fortunately, however, in the area of publication, recent developments in technology and self- and online publishing, not to mention personal websites and blogs, have brought about many desirable changes. They (to a degree, of course) have opened up several economical and efficient routes to circumvent the damaging effects of the peer review system and make one's ideas available to the public in almost any form that is desired. The authors of the past did not have peer review and had no need for it. No one peer-reviewed Plato, or Dante, or Newton, or Leibnitz. Why and how this oppressive and corrupted system developed and flourished (mostly in the 20th century) has been discussed by many authors, including me (e.g. Berezin 1996, 1998a, 2001). All too often peer review works (or at least, is intended to work) as a kind of intellectual police service, cutting off and suppressing everything which (in its view) does not conform to the established party line of the mainstream science of today (Forsdyke (1993), (2000), Feyerabend (1975), Gordon (1993), Osmond (1983), Horrobin (1990), (1996), King-Hele (1975), Kenward (1984), Kostoff (1997), Oreskes (1999)—to list just a sample of peer-review critics).

In essence, the modern business model of the science establishment is suspicious and unreceptive to new ideas, which almost always include hypothesizing and 'guess work'. All counter-claims notwithstanding, the bottom line of the highly commercialized research system of today is to support incremental research along well-established research streams in primary areas of science and technology, and to brush-off everything that seems to be 'off-line'. This is a kind of 'do and think as we do' mentality of which peer review is just one component. This, of course, does not mean that all incremental and peer review-endorsed research is useless (most certainly, there is a lot of good research in this fold), yet, many of the novel and groundbreaking research ideas are generated at the margins and fringes of mainstream science. Fortunately, the personal enthusiasm and creativity of many scientists (both professional and amateur) often makes it possible to 'sneak' many good fringe ideas into active discourse and attract broader public attention to them.

Of course, many people try their best to get around (or, perhaps, some say, 'circumvent' or 'cheat') the peer-review system. There is nothing wrong in trying to outsmart this corrupted and intellectually oppressive system. Suffice it to say that in publishing my papers I used all the 'hooks and tricks' to get around peer review and minimize its impact on my publications and ideas. I hope that such a fight on my part was largely successful (more on that in chapter 6).

Over the years I have encountered a whole spectrum of reactions to my ideas on isotopes and other matters. They ranged from true interest, healthy curiosity and even some degree of admiration, to calling me a 'heretic' and a 'crackpot' (typical name-calling for anyone who is deflecting from the mainstream). Not all such 'inputs' reached me and my guess is that a lot more was said about me behind my back. Yet, I am not at all upset by that and even see it as a kind of an honor. I am certainly in good company for that. On some occasions I was called a 'visionary', a much nicer term than crackpot, although it bears some flavor of weirdness or oddity. At one point, somebody mentioned that my papers (on isotopes in biology) have, perhaps, more ideas than I myself may realize ('your papers have, perhaps, more in them than you yourself realize'—I certainly see it as a great compliment). In this regard I like a quotation by Salvador Dali, 'When I paint and I myself don't understand what I am painting, that certainly does not mean that my images have no meaning' (as quoted by Linde Salber (Salber 2004, p 35)).

Turning back to the main course of this book, we can start with the trivial point that the existence of isotopes is one of the primary and best-known facts of physics and chemistry. Of course, 'everyone knows that' (assuming at least a basic science education). It is not so much a 'fact' of any science, but rather a primary factor in the material world. As was mentioned above, several major areas of science and technology, such as nuclear power engineering, nuclear medicine, isotopic geology (e.g. carbon dating) and a few other areas are fundamentally based on isotopes. Most of these applications make use of radioactive isotopes. The use of stable isotopes of chemical elements forms a much smaller niche, at both the technological and economic levels. For example, in the key areas of micro- and nano-technology, from information processing systems to biological and medical applications, there is (at least, at the present time) almost no use of stable isotopes (isotopic diversity). However, in view of the fact that stable isotopes of any given (poly-isotopic) element vary (least slightly) in their physical and chemical properties, there is room for a purposeful utilization of these differences. The subsequent sections of this book discuss some ideas on the use of the diversity of stable isotopes (isotopic engineering), with an emphasis on nanostructures, nano-biology and nanotechnology in general.

This book, which can also be seen as a collection of (almost independent) essays presents an attempt to build a bridge between the phenomenon of isotopicity (the fact that most chemical elements have several stable isotopes) and various areas of physics, chemistry and biology. The latter includes higher levels of biological organization that border on the domain of consciousness, spirituality and creativity. Despite the fact that the chemical properties of different isotopes of the same element are almost identical, they are not exactly the same and, what is more important, they are macroscopically distinguishable with respect to a number of physical and chemical effects. In this book I discuss a possible role for isotopic diversity in Nature as a separate and independent factor capable of playing a special role in biological, evolutionary and mental processes.

As a general remark, the author suggests that the primary philosophical quest behind the concept of isotopic biology can be formulated in the following way. If Nature is smart enough to use the diversity of chemical elements for biology (almost all the elements in the periodic table have some biological functions, including microelements), then it may look somewhat odd that Nature would omit to use such a mighty additional informationally rich resource as the diversity of stable isotopes for the structuring and functions of biological systems at all levels of evolution and complexity. The likely 'answer' to such a 'puzzle' is that, yes, Nature most likely uses it (isotopic diversity), but we have so far failed to detect this and have even (largely) failed to look at it, even at the level of a hypothesis, not to mention through targeted experimentation. One of the prime aims of this book is to draw the attention of the world's research community to this incipient research area of stable isotopicity and isotopic engineering—a direction that (with some luck) may turn out to be a newly found gold mine for physics, biology, biomedicine, material science, cognitive sciences and informational technology in a broader sense.

There are many interesting and challenging directions for isotopicity (the phenomenon of the diversity of stable isotopes), which are discussed in this book. This 'isotopicity book' explores (often in a sketchy manner) a somewhat specific area with many potential, but still largely unexplored, applications. An area which cumulatively can be called 'stable isotope engineering' has potential applications to nanoelectronics, bioengineering and the newly emerging discipline of quantum computing. Many of these ideas and hypotheses are outlined later or are at least mentioned in this book.

To return to the previously mentioned example, the main element of life, carbon, is a mixture of $^{12}$C (99%) and $^{13}$C (1%). Both are stable isotopes. One per cent of $^{13}$C may not seem a lot, but if we realize that in a human body there are billions upon billions of carbon atoms (about $10^{26}$—ten to the power of 26), this 'mere' 1% translates into a huge absolute concentration—some $10^{24}$ (a million billion billion) $^{13}$C atoms in our body.

Likewise, D (deuterium) makes up only 0.01% of all the hydrogen atoms in water (one per 10 000 atoms) and yet in absolute numbers it comes to very impressive figures. One small drop of water (say, one cubic millimeter) still contains some $10^{15}$ atoms of deuterium. Using artificial methods, D and H can be separated and the water produced (in which most hydrogen atoms are D) is known as heavy water ($D_2O$). Deuterium, and hence heavy water, is *not* radio-active, so you cannot contract radiation sickness if you bathe in it. But what would happen if you accidently drank a glass of heavy water? We will touch upon this question later.

Several other areas and aspects of stable isotopicity are considered in this book. Can it be used, for example, to build a new type of random number generator, or a new type of optical fiber, or can it be used to make microchips for quantum computers? And what about compact information storage? And do isotopes affect brain function? Or could isotopes perhaps be an essential aspect of the very mechanism of consciousness? Could there be an 'isotopic life form' out there in the outer cosmos, as an alternative to 'regular' chemistry-based biology? Can life be based on a single chemical element? These, and other questions, are discussed further in this book. As my prior experience indicates, some people may take the view that much of what has gone into this book should be classified as a 'fringe science', 'pseudo-science', or whatnot (Berezin 1996). Some stronger words are occasionally used to describe such hypothetical and speculative ideas. Some of these ideas may challenge mainstream scientific orthodoxy, as many radical ideas have before. Many such ideas are indeed dead-end offerings, but some off-line ideas have turned out to bear fruit, and for better or worse, these fruits may not always be of the kind that their original authors expected or intended.

Again, to restate what was said above (and will be repeated again below), the main conjecture proposed in this book is to view the phenomenon of isotopicity (isotopic diversity and isotopic randomness) as an additional and informationally resourceful pattern-forming capacity in Nature, which may be operational at many levels of the material world. While it may be quite subtle and might be called a 'shadow' (isotopic effects are generally weak), it is nonetheless omnipresent (isotopes are everywhere) and cumulatively its role in such realms as informational self-organization and evolution may be much more significant than has been recognized to date. It is another example of the celebrated 'butterfly effect', where seemingly small (often negligible) variations of parameters may lead to gross changes in the dynamics of a system. The author of this book has no intention of providing any definitive solutions to the discussed effects, but rather intends to exhibit them as material for further research and contemplation.

To this effect, I have always been, and I remain, on the side of those who believe that unusual and off-mainstream ideas and suggestions should be heard and studied rather than dismissed outright as 'rubbish' without even being permitted a hearing. The history of science provides many confirmations of this. It is full of examples of premature dismissals and the ridiculing of unorthodox ideas because they did not fit, as in the famous Lavoisier quotation, 'no stones can ever fall from the sky' (see section 5.6).

In my short letter 'Radical ideas should not be prematurely dismissed' (Berezin 1994), I concluded with an episode on how one of the founders of quantum physics, Niels Bohr, once replied to a teasing reporter. This reporter asked Bohr about the horseshoe nailed over the door of his summer cottage. 'Surely, Professor Bohr, you don't believe such nonsense as a horseshoe bringing luck to its owner?' To that Bohr replied, 'Of course, I don't. But they say it works even if you don't believe it'.

## References

Ades D 1982 *Dali and Surrealism* (New York: Harper and Row)

Berezin A A 1994 Radical ideas should not be prematurely dismissed *American Physical Society News* **3** 4–5

Berezin A A 1996 Mainstream and fringe scientific ideas and ultimate values *Ultimate Real. Mean.* **19** 40–49

Berezin A A 1998a The perils of centralized research funding systems *Knowl. Technol. Policy* **11** 5–26

Berezin A A 2001 Discouragement of innovation by overcompetitive research funding *Interdiscip. Sci. Rev.* **26** 97–102

Cowles F 1959 *The Case of Salvador Dali* (Boston: Little and Brown)

Dali S 1942 *The Secret Life of Salvador Dali* (New York: Dial Press)

Descharnes R and Neret G 2001 *Salvador Dali* (Cologne: Taschen)

Feyerabend P 1975 How to defend society against science *Radical Phil.* **11** 3–8

Forsdyke D R 1993 On giraffes and peer review *FASEB J.* **7** 619–21

Forsdyke D R 2000 *Tomorrow's Cures Today? How to Reform the Health Research System* (Amsterdam: Harwood Academic)

Gordon R 1993 Grant agencies versus the search for truth *Accountability Res.* **2** 297–301

Horrobin D F 1990 The philosophical basis of peer review and the suppression of innovation *J. Am. Med. Assoc.* **263** 1438–41

Horrobin D F 1996 Peer review of grant applications: a harbinger for mediocrity in clinical research? *Lancet* **348** 1293–95

Kenward M 1984 Peer review and the axe murderers *New Scientist* (31 May 1984 p 13)

King-Hele D G 1975 Truth and heresy over Earth and sky *Observatory* **95** 1–12

Kostoff R N 1997 Four factors and one criterion are key to improving peer review *Phys. Today* **50** 102–4

Kuhn T S 1962 *The Structure of Scientific Revolutions* (Chicago, IL: University of Chicago Press)

Oreskes N 1999 *The Rejection of Continental Drift* (Oxford: Oxford University Press)

Osmond D H 1983 Malice's wonderland: research funding and peer review *J. Neurobiol.* **14** 95–112

Salber L 2004 *Dali* (London: Haus)

IOP Publishing

Digital Informatics and Isotopic Biology
Self-organization and isotopically diverse systems in physics, biology and technology
**Alexander Berezin**

# Chapter 2

## Energy and information

*We are beginning to see the entire universe as a holographically interlinked network of energy and information, organically whole and self referential at all scales of its existence. We, and all things in the Universe, are non-locally connected with each other and with all other things in ways that are unfettered by the hitherto known limitations of space and time.*

Ervin Laszlo (b 1932)

'Energy' and 'information' are among the most used words across languages and cultures. We all 'understand' what these words mean. Oh, but do we? In fact, almost all the common words we use have various levels of ambiguity and contextual meaning, and, more often than not, they cannot be defined consistently at all. This chapter discusses the duality of energy and information under the angle of digital messages and the informational structure of the Universe at large.

## 2.1 The relativity and contextuality of major physical categories

*One had to be a Newton to notice that the Moon is falling, when everyone sees that it doesn't fall.*

Paul Valery (1871–1945)

Concepts, unlike specific objects, cannot be properly 'defined' in any finite and self-consistent form. Let us take a look at the five main concepts of physics (and the world in general). Let us say that they are space, time, matter, energy and information. Of these, space and time are probably the most 'fundamental'. We can imagine the world without matter and/or energy, but we can hardly imagine it without space or time. If it is a purely Platonic world of numbers, numbers as such do not 'need' either space or time for 'existence'. Take the list of prime numbers

(2,3,5,7,11,13,17,19, ... *ad infinitum*)—does it 'exist' on its own without any need of space, time, or anything else?

Of course, when a child or somebody 'uninitiated' asks us what any of these concepts mean, we usually give a lot of synonyms in reply, saying that time is the 'duration' between events, or that space is the 'extension' between objects, or something in which all objects are placed. Well, such circular 'definitions' are not much more than mere semantics and they do not move us very far towards understanding what these terms really are. Any dictionary of synonyms can do a good job here. Fortunately, though, such a lack of truly revealing definitions usually does not bother us too much (unless we are professional philosophers or theologians, perhaps). Thus, we can just use these words for the problems we are working on. As if we actually understand them.

In fact, attempts to logically define major physical categories inevitably move our quest to the realm of philosophy and metaphysics. And here, as the history of ideas demonstrates, the best we can do is to formulate open-ended questions and contemplate them rather than attempting to give any clear-cut answers. The latter is largely impossible, but, fortunately, for almost all practical purposes it is unnecessary; physicists (both experimental and theoretical) can work well with all these concepts without bothering to define them beforehand.

The situation with information or randomness is a bit different, though. While mathematics cannot do too much to help us to define (or envision) what space or time are, it (mathematics) can do a lot to help us to define the concepts of information and randomness. Here we have several operational tools, as discussed in the following sections (2.3, 2.4 and 2.5).

As has already been mentioned, while this book was written by a physicist, it is quite interdisciplinary in its coverage and approach. This, to a large degree, is influenced by my personal interests and my inclination to treat almost any physical issue I have come across under the angle of philosophy and, if you wish, metaphysics (the demarcation between philosophy and metaphysics is somewhat fuzzy and murky). As an author, I am well aware that attitudes towards philosophizing among physicists (and scientists, in general) show a great spectrum of diversity, from up-front dismissal ('not for me') to deep personal involvement and pursuit. On that note, I would like to include a marvelous quotation from the introduction to *The History of Western Philosophy* by Bertrand Russell (Russell 1945).

*Almost all the questions of most interest to speculative minds are such as science cannot answer, and the confident answers of theologians no longer seem so convincing as they did in former centuries. Is the world divided into mind and matter, and, if so, what is mind and what is matter? Is mind subject to matter, or is it possessed of independent powers? Has the Universe any unity or purpose? Is it evolving towards some goal? Are there really laws of nature, or do we believe in them only because of our innate love of order? Is man what he seems to the astronomer, a tiny lump of impure carbon and water impotently crawling on a small and unimportant planet? Or is he what he appears to Hamlet? Is he perhaps both at once? Is there a way of living that is noble and another that is base, or are all ways of living merely futile? If there is a way of living that is noble, in what does it consist, and how shall we achieve it? Must the good be eternal in order to deserve to be valued, or is it worth seeking even if the Universe is inexorably moving towards death? Is there such a thing as wisdom, or is what seems such merely the ultimate refinement of folly? To such questions no answer can be found in the laboratory. Theologies have professed to give answers, all too definite; but their very definiteness causes modern minds to view them with suspicion. The studying of these questions, if not the answering of them, is the business of philosophy.*

The above quotation has stuck with me for many years. I can practically say it by heart, and it has always been a source of much inspiration for me in many of my endeavors in science, philosophy and all other physical and metaphysical pursuits.

## 2.2 Boltzmann–Shannon informational entropy

Information as a physical category can be incorporated into the network of other physical concepts in a number of ways. In particular, the notion of information is closely related to such issues as complexity, entropy and self-organization. Two systems with the same energy content can have quite different entropies. Entropy production in an open system acts as a catalyzer of self-organizing activity and pattern formation. This is something that is symbolically implied by the very word information (in-*formation*). This often results in the generation of dissipative spatio-temporal structures.

For example, the bulk energy content of (A) an instantaneously ordered ('super-latticed') distribution of particles and (B) their 'chaotic' (random) distribution can be the same. Despite this, both phases, (A) and (B), have quite different entropies due to their different levels of organization.

Entropy, in turn, is related to the information content. The best known and the most often used relationship connecting information content and entropy is the Boltzmann–Shannon equation (Gray 2009, Shannon 1951):

$$S = -\sum p \cdot \ln(p),$$

where $S$ is the entropy of the system and $p$ is the probability of its given (quantum or classical) state $p$. Here 'ln' is the natural logarithm. The sum is taken over all states. There are also some other alternative, non-Shannonian, definitions of entropy. The major difference between energy and information is, though, not based on any unambiguous quantification scheme. Essentially, it is a matter of control and guidance rather than a measurable 'sharing'. The first (control and guidance) is actually more important for the active definition of information than the second (sharing ability). As such, information (at least sometimes) should be understood primarily in a non-additive manner, like

$$I + I = I.$$

This last (symbolic) equation means that the copying (cloning) of information does not increase its actual 'amount' (though sometimes it can increase the efficiency of its usage). Two city maps (or any number of maps, for that matter) are generally not much more useful than just one because (even if they are edited differently), they still carry (deliver) essentially the same message and bring the same informational content.

In biology there are some specifics in estimating the informational content in biological structures (Volkenstein 1977). The biological dynamics of any living system includes, apart from information and entropy in the sense of the above equation, a value factor (Turing 1952). The latter concerns how important this-or-that segment of information is for the successful implementation of the vital functions of the bio-system, be it a single organism at any level (from unicellular to higher animals and humans) or a symbiotic and interconnected larger bio-system (up to the entire biosphere, for which the Gaia metaphor is often used—see section 5.12).

## 2.3 Digital strings and normal numbers

Practically all current information storage and information processing systems operate on the principle of digital strings. That means that the information is stored and transmitted in the form of long (usually binary) strings that can be read and manipulated by the sequential processors. Any text, any picture (of any dimension or color, etc) can be coded this way. And the key question here is how to determine if a digital string carries 'real' information, or if it is just random and meaningless 'noise'. This question turns out to be a truly tricky one on all levels and calls for us to define 'randomness' and what strings can be considered 'random'. And the definition of randomness is far from trivial (Wolfram 1985); one may even, somewhat jokingly, say that the definition itself is in some way random.

For example, we say that when we throw dice or play roulette the outcome is random. But what exactly do we mean here by random? There have been many attempts to define the randomness mathematically, but all of them, when carefully thought through, leave us somewhat unsatisfied. There is the popular metaphor of the 'typing monkey', which says that a monkey typing at random on a keyboard, given unlimited time, will eventually type all of Shakespeare's plays (or any other book for that matter). Never mind that the time needed for that may be beyond imagining, many, many times greater than the alleged age of the Big Bang universe. However, the important point is that this 'monkey time', no matter how huge it is, is still *finite*.

In fact, the time needed for that can be quite easily estimated and the result can even be written compactly using tower-exponential notations, like the famous Skewes number (that is, $10^{10^{10^{34}}}$), which is often used as an example of 'super large numbers' (Skewes (1933), (1955), Knuth (1976), Berezin (1987)). While Skewes numbers are far, far greater than anything we can realistically count (e.g. the number of electrons in the Big Bang universe is 'only' $10^{90}$ or so), they are still only four-story exponents and hence far, far smaller than the other tower exponents that are used in various mathematical theorems, like, for example, the Graham number

(Ronald Graham), which has zillions upon zillions of levels and can only be written with special (arrow) notations (Knuth 1976).

The other side of this is that any unlimitedly long random number contains *all* possible information. In mathematics a truly random number is normally known under the term 'normal number'. By definition, the normal number has any sequence of digits of a given length occurring with equal probability. The next section discusses other interesting features of normal numbers.

The problem of the 'eternal records' of all possible knowledge is an interesting one. In principle, all possible knowledge is trivially coded in the set of real numbers. For example, imagine a real (irrational) number $X$ in the interval $0 < X < 1$, which is explicitly defined as

$$X = 0.12345678910111213141516171920212223 \ldots..(\text{to infinity}).$$

This number is known as Champernowne's constant. In order to construct it, all one has to do is to write all consecutive integers in a decimal notation ($N = 10$) one by one. Of course, such a process cannot be finished in practice (only in principle), but this does not affect the fact that $X$ is a rigorously defined mathematical constant. The above-written $X$ can be trivially coded by a simple computer algorithm, yet it 'contains' in it all integer numbers written in decimal form. Similar numbers can be constructed in a binary system ($N = 2$) or in a positional system with any other $N$-basis.

## 2.4 Universal and eternal records, or why pi is not exactly three

*317 is a prime, not because we think so, or because our minds are shaped in one way rather than another, but because it is, because mathematical reality is built that way.*

Godfrey Harold Hardy (1877–1947),
British mathematician and number theorist

Because any message, no matter how long, can be coded by some integer number, all possible records, all possible books, all the detailed descriptions of each possible universe (up to the level of all individual micro-states in them) are recorded in this innocent-looking number $X$. Furthermore, there are infinitely many variations of the number $X$ (for example, we can write any number of zeros (or any other digits) before starting 123... trail to get another version of $X$, like $X = 0.0000012345\ldots$, etc), or, perhaps, write at every step every next 'number' as many times as itself (which will produce a string $0.122333444455555\ldots$) and each so-written variation of $X$ still contains the full library of all possible records. There is, of course, an infinite variety of such variations.

The above-written number $X$ (whose decimal expression is infinite) is an example of the so-called normal number. An irrational number is called normal if in it all possible combinations of digits of a given length occur (asymptotically) with equal probability. While we are still lacking a rigorous proof that such glorious numbers as pi, or $e$, or $\sqrt{(2)}$, etc are normal, they almost certainly are (Wagon 1985). In fact, it is proven that the set of normal numbers is infinite (with the cardinality of continuum meaning that 'almost all' real numbers are normal). Thus, any possible message, any possible pattern and algorithm, is eternally coded within *any* normal number. And there is an infinity (actually, a continuum set, 'aleph one') of normal numbers.

Furthermore, each normal number contains any message infinitely many times. To put it vividly, the above-written number $X$ contains in it in a digitally coded form the full text of the Bible, or all of Shakespeare's plays, or any other possible book, infinitely many times. Moreover, this is true for any other normal numbers and the very 'number' of normal numbers is itself infinite. Indeed, the ideal Platonic world of numbers contains any possible book infinitely many times, probably, even continuously many times—to indicate the aleph one (continuous) set whose cardinality is next to the aleph zero countable set. The latter is the famous 'continuum hypothesis' of Georg Cantor (Dauben 1979).

In his novel *Contact* (Sagan 1986), the astronomer Carl Sagan (1934–96), describes a team of scientists who discover a peculiar, seemingly non-random, pattern among the digits of pi = 3.14159 .... This pattern appears somewhere very far in the decimal expansion of pi. Say, at some point the decimal expansion of pi has a million consecutive zeros, like ...35600000...[million]...000007923... And that the scientists interpret this as an 'eternal message' that was put there by some higher mind. But, in fact, there is nothing strange or mystical in that. If pi is a *normal number* (which it almost certainly is, even though to date there is no mathematically rigorous proof of this), then *any* sequence of digits will happen in it at some place and, in fact, infinitely many times. So, there will be a million, or a billion, or a trillion (etc) sequences of 0s, or 1s, or 7s (etc) somewhere in it and, of course, any book digitally coded in any possible code will occur in pi somewhere (and, again, infinitely many times with all possible variations). Such is the power of infinity (see section 2.5).

The number pi is, of course, a universal constant of mathematics which can be defined in many ways, not just as the ratio of the length of a circle to its diameter. For example, it can be defined by the infinite Leibniz series

$$pi/4 = 1 - 1/3 + 1/5 - 1/7 + 1/9 - 1/11 + (ad\ infinitum).$$

This very simple looking series, although it converges very slowly (there other serial expansions for pi which converge much more quickly, but they are more complicated) does, nonetheless, converge to the exact value of pi/4, even if this series as such has (seemingly) no clear relationship to circles.

To repeat, while the scientists in Sagan's novel interpret the appearance of this pattern in pi as some ultimate message from the super-mind that created the Universe ('the artist's signature'), such an interpretation breaks down upon rigorous analysis. The fact is that any (normal) number, including pi, does indeed contain any possible pattern of digits in its (decimal or other) infinite expansion, making the 'discovery' of Sagan's heroes a trivial observation related to the mathematical properties of normal numbers (yes, of course, any pattern in pi occurs infinitely many times).

That probably 'metaphysically explains' (if such a term can be used) why pi is not exactly three. Would it not be nice if it should be so? (Except that, perhaps, the Universe would be much more boring). However, pi is and will always be some strange, irrational and transcendental (and likely normal) number that places itself somewhere between 3.1 and 3.2. And no talk about 'other realities' or 'parallel universes' can be entertained here, because in *any* logically consistent realm, Pi will be exactly as it is, it is as absolutely eternal and unchangeable as the (infinite) list of prime numbers.

Thus, asking why pi is not exactly three is, perhaps, like asking why there is an infinity of prime numbers. In fact, since prime numbers become increasingly sparse as we progress along the number line, common sense may tell us that sooner or later the prime numbers will run out and there will be no more of them. However, contrary to such common sense, as early as the third century BCE, Euclid gave a neat and clear proof that there is no 'largest' prime number and there is infinity of them (his proof is so well known that any math text includes it, so there is no need for me to repeat it here).

The above comments do not, however, derail the idea of 'hidden messages' in normal numbers. On the contrary, the situation with hidden messages is, in a sense, even better than in Sagan's novel. The Platonic world of mathematics is full of them. Furthermore, this idea has enormous constructive power. The fact that real numbers (normal numbers) in their very structure contain an infinite manifold of messages and patterns of all kinds serves as the basis for all the phenomena of emergence and self-organization occurring in the Universe. This is congruent with the Pythagorean tradition of envisioning numbers (primarily integer and rational numbers) as the 'mystical foundation' of the Universe. Noting that one of the meanings of the Greek word *logos* is 'ratio', Charles Seife (Seife 2000, p 26) gave an alternative translation of John 1:1 from the Bible as 'In the beginning, there was the ratio, and the ratio was with God, and the ratio was God' (obviously, the idea of the ratio of any two integers implies the whole infinite set of numbers).

In fact, the situation with normal numbers and all possible messages (and books) may be even better than the above lines imply. In spite, as was just said, of the lack (for now, at least) of rigorous proof for the normality of any of the traditional transcendental numbers, such as pi, $e$, $\sqrt{(2)}$, etc, in 1909 Emile Borel proved the normal number theorem. It states that 'almost all' real numbers are normal in the sense that the set of exceptions (non-normal numbers) has a Lebesgue measure of zero (or, saying it in a simpler way, there are 'infinitely more' normal numbers than non-normal numbers and if you 'hit' any real number at random, the probability of 'hitting' a non-normal number is zero). So, there are plenty of good (and bad) books in the universal library hidden in the set of normal numbers (in fact, an infinity of both).

Many other examples can be brought up here. For example, Mandelbrot iterative sets show the enormous creative and diversifying power of numbers in the world. In this way, the tradition going from Pythagoras through Plato to Euclid and Diophantus and then on to Leibnitz, Cantor, Gödel, Erdos and many others, remains a guideline for many people to follow in their philosophical and metaphysical contemplations. Another interesting example here is the metaphor of the 'universal library' from the well-known short story by Jorge Louis Borges.

## 2.5  Jorge Luis Borges' *Library of Babel*

*The atoms come together in different order and position, like the letters, which, though they are few, yet, by being placed together in different ways, produce innumerable words.*

Epicurus (341–270 BCE)

This theme of infinite records has generated some interesting literature. In his story, *The Library of Babel*, Jorge Luis Borges (1899–1986) imagines the whole universe as a huge library (Borges 1998 and many other editions). To the inhabitants of the world (the 'librarians') the library appears to be infinite, having enormously long galleries of identical storage rooms with bookshelves stretching in every direction. No one knows where the galleries end or whether they end at all. This 'library of Babel' (LB) contains every possible book in every possible and imaginable language. Each book consists of an identical number of pages (410, which is of course an arbitrarily chosen number) filled with all possible permutations of letters. Therefore, all possible meaningful books are somewhere on the shelves of the LB, lost amidst an enormous number of books filled with a meaningless jumble of printed characters. There are no two identical books in the LB, but the chances of locating even a single meaningful book are very small. One has to travel many light years over the galleries of the LB to find a single meaningful book.

However, all possible books are by definition somewhere in the LB. As Borges puts it, 'Everything: the minutely detailed history of the future, the archangels' autobiographies, the faithful catalogues of the Library, thousands and thousands of false catalogues, the demonstration of the fallacy of those catalogues, the demonstration of the fallacy of the true catalogue, the Gnostic gospel of Basilides, the commentary on that gospel, the commentary on the commentary on that gospel, the true story of your death, the translation of every book in all languages, the interpolations of every book in all books' (Borges 1998).

In view of such a horrifyingly deterministic inevitability (yes, this is a mathematical certainty), any author (including me) can, theoretically, even pose a metaphysically rhetorical question: why do I have to write this-or-that book when it is 'already' on the eternal shelves of the LB? I do not have an easy answer to this conundrum, although it is, undoubtedly, a good point to ponder and reflect upon, perhaps, at some exhibition of surrealistic art. Likewise, the codes for every digitized picture—at *any* resolution—are, of course, on the selves of the LB as well. Yes, for some pictures or, perhaps, multi-dimensional images, it may take more than one 410-page book (that is the standard that Borges assigns for every book in the LB) to store the full digital code, yet *all* the books needed for this are in the LB (by the very definition of the LB).

The idea of the LB implies that all possible histories, biographies, philosophies, religious texts and scientific theories already exist, in principle, in the LB. Keeping in mind what was said above about normal numbers, we can safely say that *any* normal number will have 'inside it' the whole LB (and a lot more) and, furthermore, the LB will be there *infinitely many times* in all possible variations. This is because, while the LB is enormously huge, it is still of finite size. And no finite number can beat infinity. Thus, Champernowne's constant has the LB in itself, and so does pi (assuming it is normal, which is almost certainly so), infinitely many times.

It is possible to estimate that all the permutations of several million characters (typical book length) amount to some $T(3) = 10^{10^{10}}$ (10 to the power of 10 billion) possibilities (Bloch 2008). Such numbers are called tower-exponential numbers. No matter how high the tower of 10 is, the number it represents is always finite (which is different from normal numbers, which are all infinitely long strings of digits). Almost all integer numbers are larger than any given one we can think of. With any specific number we are always infinitely far from the 'true infinity' (Knuth 1976). Thus, while the LB is much larger than the observable universe (which is about ten billion light years across (that is 'only' $10^{10} = T(2)$), it is eventually finite (it becomes infinite if we allow books with an unlimited number of pages).

The above example, by the way, shows an *enormous* difference between $T(3) = 10^{10^{10}}$ (that is, the size of the LB) and $T(2) = 10^{10}$ (the size of 'our' Big Bang universe, BBU). While at first glance $T(2)$ and $T(3)$ may appear innocently close (there is not much difference between 2 and 3 in the world of tower exponents), these two items are fantastically different. In fact, their ratio is $T(3)/T(2) = 10^{9,999,999,990}$, which is almost as large as $T(3)$ itself. The above ratio shows how much greater the LB is than our BBU.

But there is another problem with the LB, apart from its size. It not only contains all true knowledge and facts, but also all that is false. There is no outside tool that (even in principle) could help us to separate the true from the false (except for trivial statements like $2 + 2 = 5$, the falseness of which is obvious). This lack of a truth criterion for the LB is conceptually akin to Gödel's undecidability theorem, which demonstrates the existence of unprovable statements. Another analogy is the so-called 'halting problem' for the universal Turing computer. The latter is a theorem that demonstrates that it is impossible, even in principle, to design an algorithm which can determine in a finite number of steps whether any given computer program will eventually stop (halt).

For Borges, the notion of the LB of all possible books is a metaphor for the Universe itself. By definition, the set of all possible books has the detailed history of all possible worlds written in it, even if by 'history' we mean the detailed account of all the microscopic states of each mini-verse. Of course, it takes many, many volumes of the LB to write down the entire microscopic history of any particular universe (mini-verse), but never mind, it is just fine, because all the required volumes are somewhere in the LB anyway (although any two 'consecutive' volumes may be light-years apart, sitting on the shelves of LB).

In other words, in view of the above estimate for the number of microscopic states of each mini-verse (about 10E10E10 = $10^{10^{10}}$ states), it may take an enormous number of volumes on the shelves of the LB to record the full history of even one particular mini-verse; yet the total number of such books (and all the books in the LB) is finite.

# References

Berezin A A 1987 Super super large numbers *J. Recreat. Math.* **19** 142–143

Borges J L 1998 *Collected Fictions* (New York: Viking)

Bloch W G 2008 *The Unimaginable Mathematics of Borges' Library of Babel* (Oxford: Oxford University Press)

Dauben J W 1979 *Georg Cantor: His Mathematics and Philosophy of Infinite* (Princeton, NJ: Princeton University Press)

Gray R M 2009 *Entropy and Information Theory* (New York: Springer)

Knuth D E 1976 Mathematics and computer science: coping with finiteness *Science* **194** 1235–42

Russell B 1945 *A History of Western Philosophy* (London: Unwin)

Sagan C 1986 *Contact* (New York: Pocket)

Seife C 2000 *Zero: The Biography of a Dangerous Idea* (New York: Penguin)

Shannon C E 1951 Prediction of entropy of printed English *Bell Syst. Tech. J.* **30** 50–64

Skewes S 1933 On the difference between $\pi$ (x)-Li(x)' *J. London Math. Soc.* **8** 277–283

Skewes S 1955 On the difference between $\pi$ (x)-Li(x) II *Proc. London Math. Soc.* **5** 48–70

Turing A M 1952 The chemical basis of morphogenesis *Phil. Trans. R. Soc.* B **237** 37–72

Volkenstein M V 1977 Amount and value of information in biology *Found. Phys.* **7** 97–109

Wagon S 1985 Is pi normal? *Math. Intelligencer* **7** 65–67

Wolfram S 1985 Origins of randomness in physical systems *Phys. Rev. Lett.* **55** 449

IOP Publishing

Digital Informatics and Isotopic Biology
Self-organization and isotopically diverse systems in physics, biology and technology
**Alexander Berezin**

# Chapter 3

# Chaos and self-organization in random systems

*Nature does nothing uselessly*

Aristotle, Politics

One of the primary features of our perceptive facilities is that we perceive almost everything in terms of dichotomies, or, in plain language—as opposites. Day and night, hot and cold, life and death, love and hate, good and evil—this play of opposites permeates our language and (to a large degree) our very lives. And when we descend to the world of exact sciences, primarily physics, we see much of the same: motion and rest, waves and particles, positive and negative (charges), attraction and repulsion, etc. And a mighty and highly useful dichotomy has emerged in recent decades (although some glimpses of it were recognized much earlier). This is the dichotomy of order and disorder, or (in a somewhat different wording) the dichotomy of chaos and self-organization.

In this chapter I discuss some ideas related to randomness and order, and chaos and self-organization in systems of atoms; interactions and configurations in electrostatic systems; and structural phase transitions in systems of particles that interact via Coulombic and also non-Coulombic interaction. Some ideas from Rene Thom's catastrophe theory will be used for this discussion. Some of the work that I did in these areas is connected to the ideas of isotopicity that will be discussed further in chapters 4 and 5.

## 3.1 Dichotomy of order and chaos

*We adore chaos because we love to produce order.*

Maurits Escher (1898–1972)

doi:10.1088/978-0-7503-1293-6ch3

The modern theory of chaos emerged largely in the 1960s and 1970s with the work of Edward Lorentz (weather predictions), Mitchell Feigenbaum (iterative logistic maps and bifurcation cascades), Per Bak (self-organized criticality, SOC) and many others, although some of its ideas appeared much earlier (e.g. in the work of Henri Poincaré in the 1880s on non-periodic orbits in the three-body problem). In the realm of numbers and digital strings the dichotomy of chaos and order approaches tricky issues of information theory, such as the informational content of random sequences, algorithmic complexity, etc. Much of this calls for some reflections on what can be called 'metaphysical questions' rather than yes-or-no types of answers. For example, we could inquire for eternity as to whether the digits of pi are random, or chaotic, or organized in some order. Even if it can be proven rigorously that pi is a normal number (see sections 2.3 and 2.4), there is still a dichotomy looming: digits of normal numbers are by definition random, yet there are deterministic algorithms that can be used to print out any number of digits of pi up to (theoretically) infinity.

As for the main theme of this book (isotopic randomness), the links between isotopicity and chaos are numerous, and some are strong, while some are subtle. As was mentioned above, throughout this book I will indicate some possible implications of the phenomenon of isotopicity—the natural diversity of stable isotopes—for fundamental informationally related phenomena. The latter include such areas as spontaneous self-organization in Nature, as well my offerings on the possible role(s) of isotopically related quantum effects in the physical foundations of consciousness and creativity. Isotopicity greatly diversifies most otherwise almost identical chemically defined systems (large molecules, crystals, etc) and enables them to be highly individualized systems capable of an enormous number of inner states.

Chaos theory has introduced (or revived into active use) a number of concepts and heuristic models. One concept of a particular relevance to isotopicity is the notion of the strange attractor. This theoretical construct describes the trajectory of a 'phase point' into $N$-dimensional phase space. This author has extended this concept to depict the topology of the trajectory linking discrete points in isotopic phase space (Berezin 1991). Each point represents a combination of the physical characteristics of a particular chemical compound for a fixed combination of stable isotopes.

## 3.2 Nature's quest for patterns

*A mathematician, like a painter or poet, is a maker of patterns. If his patterns are more permanent than theirs, it is because they are made with ideas.*
Godfrey Harold Hardy (1877–1947),
British mathematician and number theorist

Philosophical (or metaphysical—depending of how one puts it) reflections on isotopicity may lead to the following inferences. Isotopicity provides an additional (parallel and/or alternative) route for Nature to satisfy its alleged quest for patterns and gives an independent level of freedom within the chemical structure. Isotopic freedom faces only quite minimal constraints, imposed by other levels of organization (hetero-chemical, biodynamical, etc). The dichotomy between complexity (the trend to form rich, non-repetitive and multi-scale patterns) and simplicity (Nature's economy, following the principle of Ockham's razor) can be seen in the realm of isotopic freedom. Therefore, I suggest that isotopicity, which is currently a somewhat overlooked facet of Nature's diversity, deserves greater attention and further investigation.

In terms of how Nature creates its patterns, one can draw a useful analogy from the art of computer simulations. The realistic appearance of computer-generated landscapes is but one argument to suggest that complex natural systems can be efficiently coded by relatively short algorithms. The remarkable persistence of the 'Babylonian library' (BL, section 2.5) idea of all possible books (including pattern-forming instructions), from the ideal Platonic world of forms to the essays of Jorge Luis Borges, illustrates our search for a unique principle to encompass cosmo-genesis, emergence and self-organization ('for deriving all from nothing there suffices a single principle', as a quotation attributed to G W Leibnitz has it). The often noticed 'unreasonable effectiveness of mathematics' (Wigner) recasts the Pythagorean 'all things are made of numbers'.

At various physical levels, such a single universal principle (Wigner's law) acquires numerous specific reincarnations, as has been attested to by many authors from a variety of positions and often in different ways and forms (e.g. Casti (1990), Lloyd (2006), Pickover (1995), (2001), (2007), Penrose (1994), Plichta (1997), Tipler (1994), Wilczek (1999)). The general scenario proceeds from the delivery of particular abstract patterns to a specific level of their implementation. Take, for example, our carbon-based life forms. How did they originate? Their origin, including its sentient level (consciousness), calls for some connecting agent positioned (in a metaphysical sense) between the BL of all patterns and specific biochemical structures. A possible candidate is the isotopic diversity of chemical elements.

Furthermore, 'higher alephs' (Cantor's sets) also become 'available' through integer truncations to finite digital strings (e.g. a string of digits of pi of any finite length). Likewise, patterns of the (mega-) universe at tower-exponential scales (Dyson 1979) may also be engaged. For that, the Platonist axiom that there *is* an infinite Euclidean embedding space of any dimensionality, suggests that *some* kinds of structures and objects are occurring at *any* scale of the mega-universe (e.g. Tegmark 2014, Yourgrau 1999).

The potential convenience of isotopicity for the digitization of informational dynamics (isotopes are discrete entities and hence integers) makes it possible for them to use Gödel-type numbering involving integer powers of primes (Casti 1990). Informationally rich, super-long tower-exponential integers (Knuth 1976) can be 'downloaded' from BL through quasi-fractal trees of tower exponents of primes ($N(n) = p1^\wedge p2^\wedge...^\wedge pn$). Every combination of primes defines a specific 'integer empire'. By this (peculiar, of course) symbolic term I mean the set of all integers between $N(n)$ and its next neighbor, which is $N(n+1)$. Of course, because of the very mathematics of tower exponents (Knuth (1976), Berezin (1987), (1998), (2004)), $N(n+1)$ is immensely greater than $N(n)$. Each so defined integer empire has a unique pattern of primes. These patterns may later serve (in a metaphysical sense) as a blueprint for self-organizational dynamics. Isotopic clusters can implement such recording in a microscopically compact way (perhaps, at the nanoscale level). Supplementary to brain and neural functioning, isotopicity may play some role in quasi-biological activity in solutions and contribute to the memory transfer aspects of (what is claimed to be) water memory and 'homeopathic' effects (more on that in chapter 5). Another example of quasi-biological activity may be the plasmatic DNA-like structures mentioned below in section 5.17 (e.g. Tsytovich *et al* 2007).

Almost any heterogeneous system can form patterns of a varying degree of complexity. The degree of complexity can be very low to very high. For example, a symmetrical pattern, like the periodical combination ABCABCABCABCABC(. . .) of the same unit (ABC) has a low degree of complexity, while a-periodical (non-periodical) strings like ACBAABCACBABBCBCA(. . .) have much higher complexity. Likewise, a periodically repeated string of all the letters of, say, the Latin alphabet (ABCDEF... XYZ) indefinitely repeated has a much lower degree of complexity than any ('real', that is 'meaningful') book which uses the same set of 26 letters (adding capital letters and common punctuation marks will bring the total number of symbols to some 70 or 80—that is still a pretty small set of characters). This problem was discussed by Jorge Luis Borges in his *Library of Babel*. In that sense, systems with isotopic diversity provide a rich playground for complexity emergence and pattern evolution, as the number of spatial combinations for isotopes is truly enormous.

## 3.3 Pythagoras, 'everything is a number' and modern physics

Together with many others, I view information as a category of physics and/or mathematics. This is somewhat different from the common use (abuse?) of the word 'information' in everyday language and life. As everyone knows, much of what passes for information in the public sphere (press, public speeches, especially by politicians, etc) is, in fact, *dis*- or *mis*-information.

Pythagoras (~570–495 BCE) is mostly known for his Pythagorean theorem for rectangular triangles (Sangalli 2006, Giordano 2011, Martinez 2012). However, on the philosophical side his main idea concerned the fundamental role of numbers in the Universe. At the core, he maintained that the ultimate reality of the Universe is 'number', a view which later flourished as a major philosophical theme. While Plato (427–347 BC) cannot be called a direct disciple of Pythagoras, many of his ideas fall into the same fold. And as such, Pythagorean tradition (Pythagoreanism), which posits that 'everything is number', finds its uptake in the notion of the ideal Platonic world (IPW) of numbers and forms. And that remains an ongoing philosophical discourse up to modern times.

Such modern Platonists as Georg Cantor, Kurt Gödel and Roger Penrose (with apologies to the many I have not mentioned) follow (in their writings) this Pythagorean–Platonic tradition and develop this line of thinking within the high standards of modern mathematics, physics, mathematical logic and information theory. As the author of this book, I see myself within the same tradition, as should be obvious to anyone who bothers to read these pages. And in the visual arts, directions such as surrealism (e.g. Salvador Dali or Yves Tanguy) express a good deal of the Pythagorean–Platonic spirit.

In stressing 'number' as the foundation of the Universe, it may be said that Pythagoras had some kind of a pre-cognition of our digital age, when almost all of our information and communication technology rests on the use of digital strings. Digitization is everywhere nowadays. Nothing goes without digits. And quantum physics with its ideas of discreteness and quantum states (eigenstates) is also pretty much on the side of this digital worldview. As the Nobel-prize physicist Frank Wilczek mentioned, 'classical physics is profoundly anti-Pythagorean' (Wilczek 1999). At the same time, *quantum* physics and general relativity (the theory of gravitation) open the way to what Wilczek calls 'modern Pythagoreanism'. The combination of the Newtonian constant of gravity ($G$), Planck's constant ($h$) and the velocity of light ($c$) allows us to construct a fundamental unit of length, called the Planck length. It is defined as $\sqrt{(Gh/c)}$ and its numerical value is about $10^{-35}$ m, a far smaller length than can be imagined by our senses.

The Planck length is a fantastically small length, 35 orders of magnitude smaller than our human dimensions. The size of the (Big Bang) universe (and we, so far, do not know what may lie beyond it) is estimated as $10^{26}$ m or, equivalently, $10^{35}$ nm (if you put three or four atoms together that is a nanometer). So, apart from a factor of three or four, the Planck length is as small in comparison with an atom as an atom is in comparison with the Universe. And yet, this length, the Planck length, is among the fundamental units of physics and it is often used in quantum physics and cosmology.

Combining the Planck length with another fundamental length in quantum physics, the Bohr radius (which is 0.0529 nm, or 0.529 Å—the radius of the hydrogen atom) we obtain, in the spirit of Pythagoras, a dimensionless number. The ratio of the Bohr radius to the Planck length is of the order of $10^{26}$, an immensely large number. However, it is dimensionless (a pure number), even if it cannot be defined with such precision that we can determine whether it is an integer number or not (most likely it is not). Other similar dimensionless ratios can be obtained by using other fundamental lengths in physics, for example, the Compton length. Some units other than length (for example, time or energy) can also lead to a variety of dimensionless ratios, as with the famous fine structure constant 1/137. All this appears to be in the spirit of the Pythagorean idea of reducing everything to numbers (or rather explaining everything by numbers).

However, I have a somewhat cautious comment to make about this line of activity, which Wilczek calls the Pythagoras–Planck program. There is no real guarantee that the fundamental constants of physics do not change (albeit slowly) with time or do not depend on the position of the observer in the (Big Bang) universe. Maybe so, maybe not, but anyone thinking about these issues will most likely run into such concerns. Yet, for my own narrative, which is focused along the paradigm of isotopicity, this objection does not appear to be of critical importance.

## 3.4 Digital Pythagoreanism and the omega number

*It's much more fun to be contradicted than to be ignored.*
Freeman Dyson (b 1923)

The crossover between isotopicity and Pythagoreanism lies, in my view, in the digital nature of isotopes, which can be counted individually by numbers. The string of DNA containing a chain of carbon atoms reads like a digital string of the type 1001010111001010, etc (for simplicity, let us assume that we have a $^{13}$C-enriched carbon chain with half of all the carbon atoms being $^{13}$C). And this is the same idea of digital-like informational patterns that I present in this book when talking about the memory effect in water and homeopathy (Berezin 1990, 1994a). In chapter 5, I advance these ideas into other outlets of physics and/or metaphysics, however one might like to describe it.

For example, we say that when we throw dice or play roulette the outcome is random. But what exactly do we mean here by random? There have been many attempts to define the randomness mathematically, but all of them, when carefully thought through, leave us somewhat unsatisfied. There is the popular metaphor of the 'typing monkey', which says that a monkey typing at random on a keyboard, given unlimited time, will eventually type all of Shakespeare's plays (or any other book for that matter). Never mind that the time needed for that may be beyond imagining, many, many times greater than the alleged age of the Big Bang universe. However, the important point is that this 'monkey time', no matter how huge it is, is still finite. In other words, any unlimitedly long random number contains *all* possible information. In mathematics a truly random number is normally known under the term 'normal number', as was discussed in chapter 2.

To further these ideas, the mathematician Gregory Chaitin introduced the so-called *omega number* into the theory of randomness and (un)computable functions. It is related to the halting probability of computer programs, something that was initiated by Alan Turing (1912–54). Omega number is a generic term and it can be defined in many ways, so it is not unique like, for example, pi (3.14159…). The omega number is a real transcendental number that can be defined, yet there is no algorithm that can compute it. It sounds paradoxical, and it is, being somewhat akin to Gödel's undecidability theorem, yet, as I have already stated, I like the paradoxes and antinomies.

While an absolute ('true') randomness is perhaps Gödel-undecidable (undefinable in a closed form), in practice it is often measured by the information content needed to describe (explicit or implicit) patterns (e.g. Shannon entropy). Like stars in the sky (are they 'random' or do they form 'constellations'?), stable isotopes in crystals form patterns. Informational patterning through correlated neutron tunneling (CNT, Berezin 1992a, see sections 5.10 and 5.11) indicates the limitations of traditional statistics based on the 'democratic' (ergodic) counting of available configurations. Isotopic systems of finite size may be governed by an informational attractor with infinite Shannon entropy (of countable or uncountable cardinality, in the sense of Cantor's alephs).

Furthermore, as with the omega number, the true structure of this informational attractor may even be ultimately unknowable. Such unknowability, however, does not preclude the said attractor from possessing a pattern-forming capacity. Infinite sets are not confined to classical probabilities, e.g. the infinite set of all integers can be split on the infinite set of infinite sets, each containing an infinite set of primes and only one composite. This is the 'statistical inversion' discussed by David Lewis (Lewis 1986, Berezin 2015). Thus, the fact that CNT occupies only a tiny (tower-exponentially small) fraction of Hilbert space can be overridden by some nonergodic attractor implying Aristotelian 'final causation' (*causa finalis*). Likewise, correlated radioactive decays imply a possibility of strong (laser-like) departures from standard exponential decay law. (In his philosophy Aristotle defined four types of causes: the *material cause*, 'that out of which', e.g. the bronze of a statue; the *formal cause*, 'the form', 'the account of what-it-is-to-be', e.g. the shape of a statue; the *efficient cause*, 'the primary source of the change or rest', e.g. the artisan, the art of bronze-casting the statue, the man who gives advice, the father of the child; and the *final cause*, 'the end, that for the sake of which a thing is done', e.g. health is the end of walking, losing weight, purging, drugs and surgical tools).

A few more words about the 'paradox of infinity'. Strange as it may appear at first glance (and it may even seem counter-logical), theoretically and conceptually it is easier to work with infinite ideal systems than with real and finite. Similar paradoxes of infinity are known in mathematics, for example, many functions (e.g. Gaussian exponents) are easier to integrate from zero to infinity than to a finite limit. Mathematically, it is far easier to describe all integer numbers $(1,2,3,4,5,6,7,\ldots)$ than a smaller subset of prime numbers $(2,3,5,7,11,13,17,19,23,\ldots)$. The program (algorithm) for the set of all numbers is trivial, while the program which types out only prime numbers is much more complicated. And in the case of crystals, an infinite lattice model avoids the necessity of dealing with boundary conditions at the surfaces—the latter makes the mathematical analysis of finite structures with boundaries much more complex than that for infinite periodic structures.

## 3.5 Universal emergence and the 'Platonic pressure effect'

*Two possibilities exist: either we are alone in the universe or we are not. Both are equally terrifying.*

Arthur C Clarke (1917–2008)

Throughout this book I indicate some possible implications of the phenomenon of isotopicity—the natural diversity of stable isotopes—for fundamental informationally related phenomena, including spontaneous self-organization in Nature, consciousness and creativity. Isotopicity greatly diversifies most otherwise almost identical chemically defined systems (large molecules, crystals, etc) and enables them to be highly individualized systems capable of an enormous number of inner states.

The first (in fact, the key) idea here is 'Nature's quest for patterns'. Philosophical (or metaphysical—depending of how one puts it) reflections on isotopicity may lead to the following inferences. Isotopicity provides a route for Nature to satisfy its alleged quest for patterns and gives an independent level of 'freedom within the chemical structure'. Isotopic freedom faces only quite minimal constraints, imposed by other levels of organization (hetero-chemical, biodynamical, etc). The dichotomy between complexity (the trend to form rich, non-repetitive and multi-scale patterns) and simplicity (Nature's economy, following the principle of Ockham's razor) can be seen in the realm of isotopic freedom. Therefore, I suggest that isotopicity, which is currently a somewhat overlooked facet of Nature's diversity, deserves greater attention and further investigation.

In terms of how Nature creates its patterns one can draw a useful analogy from the art of computer simulations. The realistic appearance of computer-generated landscapes is but one argument to suggest that complex natural systems can be efficiently coded by relatively short algorithms. The remarkable persistence of the 'Babylonian Library' (BL) idea of all possible books (including pattern-forming instructions), from the ideal Platonic world of forms to the essays of Jorge Luis Borges (Borges 1998, Bloch, 2008) illustrates our search for a unique principle to encompass cosmogenesis, emergence and self-organization ('for deriving all from nothing there suffices a single principle'—G W Leibnitz).

In this regard, the often noticed 'unreasonable effectiveness of mathematics' (Wigner) recasts the Pythagorean 'all things are made of numbers'. At various physical levels, such a single universal principle acquires numerous specific reincarnations. The general scenario proceeds from the delivery of particular abstract patterns to a specific level of their implementation. Take, for example, our carbon-based life forms. How did they originate? Their origin, including the sentient level (consciousness), calls for some connecting agent positioned (in a metaphysical sense) between the BL of all patterns and specific biochemical structures. A possible candidate is the isotopic diversity of chemical elements.

For example, the flexibility in the positions of $^{12}$C and $^{13}$C atoms (different nuclear spins) in a DNA string opens up the possibility of messages that are 'overwritten above' (or are independent of) the level of chemical diversity (an analogy is the so-called 'subliminal messages' that some claim can be added to music). Isotopic diversity within chemically fixed crystalline structures admits 'freedom within determinism'. In spite of the fact that isotopic effects are generally subtle (energetically weak), they can, in the spirit of the butterfly effect, informationally amplify themselves to levels that have profound consequences for a system. Because of the huge number of atoms involved, isotopic distributions may act as intermediaries between BL and the physico-chemical level of biological functioning. Isotopicity then acts as a 'detector' of universal BL patterns, connecting them to the atomic–molecular level. Once such a connection is established (through isotopicity and/or other mechanisms), the entire aleph naught of BL content (coded in the countable infinitude of all possible digital strings) becomes 'available' to foster self-organizational dynamics.

## 3.6 Cantor, Gödel and 'ultimate issues'

*I am convinced of the afterlife, independent of theology. If the world is rationally constructed, there must be an afterlife.*

Kurt Gödel (1906–1978)

Here the idea that I am talking about can be spelled out as the 'pressures from mathematical infinity to create Nature'. Can mathematics (pure numbers; abstraction) re-incarnate itself as physical reality? Or, perhaps, it is some kind of a reshuffling of a central idea of most religions—God (something 'immaterial') creating the real world of matter and energy? 'Mathematics as a God', so to speak.

Well, in facing the 'eternal' problems, some of us may come to search for what may be called 'metaphysical invariants'. Probably, one of the best-known trails in this direction can be found in the area of pure mathematics. Here we are dealing with the issue of metaphysical pressures, of the 'desire' for 'embodiment' from the ideal Platonic world (Berezin 1998). Let me first explain the terminology and introduce the tools for the discussion. Two names seem particularly pertinent in this context. Both are great mathematicians with strong philosophical inclinations. The first is Georg Cantor (1845–1918) and the second is Kurt Gödel (1906–78). Cantor is mostly known for his ideas on the hierarchies of infinities, and Gödel for his incompleteness theorem. Both of these developments are broadly perceived as outstanding intellectual achievements.

Cantor's studies of the structure of infinities was seen by some of his contemporaries as being on the verge of blasphemy. Leopold Kronecker (a great mathematician in his own right) considered Cantor to be a scientific charlatan, a renegade, a 'corrupter of youth'. Another great scientist, Henri Poincaré (one of the precursors of modern chaos theory) thought that set theory and Cantor's transfinite numbers represented a grave mathematical malady, a pathological illness that could one day be cured (Dauben 1979). Of course, not everyone was that negative about Cantor's ideas. Bertrand Russell described him as one of the greatest intellects of the nineteenth century (Russell 1989). David Hilbert ('Hilbert space' in quantum physics) believed that Cantor had created a 'new paradise' for mathematicians (Dauben 1977, 1979) and maintained the highest regard for Cantor.

Furthermore, Cantor's ideas made a fundamental impact on thinking about the eternal issues, an impact that is probably still not fully appreciated. As Dauben said about Cantor's views: 'he summarized the position commonly encountered in the seventeenth century: that the number could only be predicated of the finite. The infinite, or Absolute, in this view belonged uniquely to God. Uniquely predicated, it was also beyond determination, since once determined, the Absolute could no longer be regarded as infinite, but was necessarily finite by definition. Cantor's inquisitive "how infinite" was an impossible question. To minds like Spinoza and Leibnitz, the infinite in this absolute sense was incomprehensible, as was God, and therefore any attempt to assign a basis for determining magnitudes other than merely potential ones was predestined to fail' (Dauben 1979, p 123).

Cantor's major accomplishment was, perhaps, a clear introduction of the idea of the cardinality of infinite sets. For example, a set of all integers (1,2,3,4,5,6,7,...) has the so-called aleph zero cardinality (countable set). This is, so to speak, the lowest level of (imaginable) infinity. Any (infinite) set which can be put in a one-to-one correspondence with the set of integers has exactly the same cardinality, aleph zero. For example, no matter how counterintuitive it may appear at first glance, the number of all integer numbers and the number of squares is the same; there are not fewer squares than integers. It can be seen from the one-to-one correspondence between 'all' integers and 'just squares' that the pairing of all integers and their squares (and such a pairing can go to infinity) clearly demonstrates that both sets have the same 'size' (same cardinality): 1–1, 2–4, 3–9, 4–16, 5–25, 6–36, 7–49, 8–64, 9–81, 10–100, 11–121, 12–144 (*ad infinitum*).

In other words, strings of all integers (1,2,3,4,5,...) and the string of only squares (1,4,9,16, 25,...) have the same 'length' (infinite, of course), and each square has a fixed 'partner'—an integer of which it is a square. Likewise, the number of primes (or their squares, or their millionth powers, etc) is exactly the same as the number of all integers, and so is the number of all rational numbers (ratio $p/q$ on any two integers).

Because primes become progressively more rare among integers, common sense tells us that there are infinitely fewer primes than composites (because, asymptotically, almost all integers are composites). If we were to choose any integer at random among the infinity of all integers, the probability of its being prime tends to zero. This is indeed so if we review all integers in the order of their natural appearance (1,2,3,4,5,...). However, David Lewis in his book *On the Plurality of Worlds* gives a simple example to illustrate the paradox of infinity. Let us take the (infinite) set of integers and rearrange it in the form of two-dimensional array (Lewis 1986, p 119) in which the first column has all integer numbers (1,2,3,4,5,6,7,8,9,10,11,...) and to the right of it there are all primes arranged in a zig-zag pattern as shown in figure 3.1. For each integer in the first column there are *infinitely many* primes in the corresponding line. There is no repetition in this infinite table and every prime is listed in it.

Therefore, each horizontal line is infinite and has an infinite number of primes. All integers occur in this table, each appears just once. For example, any composite integer will eventually be reached if we follow down the first column. And every prime will be found somewhere on one of the horizontal lines. Yet, we see that for each row, the primes outnumber composites by infinity to one. According to the image evoked by such a table, primes predominate among all integers, in spite of the fact that 'we know' that the probability of 'hitting' a prime number at random is zero (if we use all numbers with the same probability). Lewis uses the same logic to demonstrate that the 'improbable' worlds whose (formal) probability is vanishingly small (such as worlds containing intelligent life) not only can exist, but (in some sense) can predominate among all possible worlds.

Likewise, the number of all real numbers (including such numbers as the square root of 2, or pi, or *e*, etc) has a higher 'rank of infinity', known as aleph one cardinality. The number of all points on a line (an infinite line or just a finite segment

## PARADOX OF INFINITY

| 1 | (1) | 2 | 11 | 13 | 43 | 47 | 103 | 107 | 193 | 197 | 313 | 317 | 463 ... |
|---|-----|---|----|----|----|----|-----|-----|-----|-----|-----|-----|---------|
| 2 | 3 | 7 | 17 | 41 | 53 | 101 | 109 | 191 | 199 | 311 | 331 | 461 ........ | |
| 3 | 5 | 19 | 37 | 59 | 97 | 113 | 181 | 211 | 307 | 337 | 457 ............. | | |
| 4 | 23 | 31 | 61 | 89 | 127 | 179 | 223 | 293 | 347 | 449 ................. | | | |
| 5 | 29 | 67 | 83 | 131 | 173 | 227 | 283 | 349 | 443 ...................... | | | | |
| 6 | 71 | 79 | 137 | 167 | 229 | 281 | 353 | 439 ........................... | | | | | |
| 7 | 73 | 139 | 163 | 233 | 277 | 359 | 433 ................................ | | | | | | |
| 8 | 149 | 157 | 239 | 271 | 367 | 431 ..................................... | | | | | | | |
| 9 | 151 | 241 | 269 | 373 | 421 .......................................... | | | | | | | | |
| 10 | 251 | 263 | 379 | 419 ............................................... | | | | | | | | | |
| 11 | 257 | 383 | 409 ................................................... | | | | | | | | | | |
| 12 | 389 | 401 ....................................................... | | | | | | | | | | | |
| 13 | 397 ........................................................... | | | | | | | | | | | | |

........................................................................

**Figure 3.1.** *More primes than all integers?* This diagram, adopted from David Lewis (*On the Plurality of Worlds*, 1986), illustrates the 'Paradox of Infinity'. Here the *consecutive* prime numbers are arranged in the infinite zig-zag table in which all the lines are numbered by *consecutive* integers (zig-zag lines connecting consecutive primes are omitted from this diagram—it's easy to do it yourself—just go prime-by-by prime: 2, 3, 5, 7, 11, 13, 17, ...). This diagram creates the appearance that there are infinitely many more primes than integers! (All lines and columns are supposed to run to infinity due to the infinite number of primes available in the Infinite Platonic World.)

of it; this makes no difference), the points on a plane, and the points of three-dimensional (or any $N$-dimensional) space have the same power ('cardinality') of aleph one. Cantor showed that there 'is'[1] an infinite hierarchy of infinities themselves. He also constructed some examples.

The above-mentioned 'metaphysical pressure' of the ideal Platonic world (IPW) or the 'embodiment' in a form of a 'material world' (or what we perceive as such), or, in other words, 'the desire' of the 'numerological patterns' for tangible manifestation ('embodiment') is the crux of my argument. In one of my papers (Berezin 1998) I introduced the term 'Platonic pressure effect' to describe the above (metaphysical) idea. This idea resonates to some degree with what John Archibald Wheeler called the 'it from bit' concept (Wheeler 1990, Wilczek 1999).

According to Wheeler, it from bit symbolizes the idea that every item of the physical world has at its bottom—at a very deep bottom, in most instances, at least—an immaterial source and explanation (see section 3.7). This is what we call 'reality' and it arises in the last analysis from the posing of yes–no questions and the

---

[1] I put 'is' into quotation marks to indicate that this relates to 'existence' in the ideal Platonic world of mathematics.

registering of equipment-evoked responses. In short, the message here is that all things that we call 'physical' are, in fact, information-theoretic in origin and this is the meaning of the 'participatory universe', as Wheeler calls it.

To restate the above argument using some anthropomorphic expressions, we can say that this metaphysical pressure results from the 'urge' of (ideal) numerical patterns to find their manifestable existence at the level of individualization and their (self)isolation from other competing patterns. However, it should not be understood that such an embodiment in any way affects the metaphysical status of numbers (as if promoting them from the category of *potentialia* to that of *actualia*). On the contrary, the ever-the-same absolute nature of numbers implies their metaphysical immunity from all the activities that the numerical patterns can catalyze (e.g. cosmological emergence, self-organization, etc).

This Platonic pressure arises by a sheer virtue of the fact that the entire infinite pattern of integer numbers and all algorithmically derivable sub-patterns (e.g. pattern of prime numbers) are instantaneously available as a 'free lunch' at any instant/point of space–time (we mean here space–time of any dimensionality and hierarchical level, not just our ordinary four-dimensional Einsteinian space–time). This infinitely rich pattern of aleph naught (countable set) acts as an 'independent' physical effect, which directly generates the physical world 'out of nothing'.

While I am unaware if the term 'Platonic pressure' was used earlier, I shall refrain (as with all the other terms I use in this book) from claiming that I originated this idea. In fact, the entire Pythagorean–Platonic tradition can be to a large degree interpreted in this vein. Furthermore, the metaphysical primacy of integer numbers (as expressed in the Platonic pressure effect) is not necessarily remote from immediate experiences we can refer to. There are numerous examples of pattern formation governed by some simple iterative laws involving specific sets of integers. For example, the spirals of sunflowers follow the pattern of Fibonacci numbers. The peculiar lagoons of the Mandelbrot set are the result of the iteration of equations where all the seed numbers are truncated to rational numbers (all computer inputs are always rational numbers).

This certainly does not mean that the appeal to Platonic pressure effects and the eternality of integer numbers immediately obliterates the most fundamental question of philosophy, which was formulated by Martin Heidegger as 'Why is there a universe?' (Hersh 1995). Why is there something rather than nothing? Even a claim of absolute priority for integer numbers still leaves the question of why there are integers in the first place unanswered. Here we have little recourse other than to take numbers for granted. However, making the minimal concession of admitting the existence of 'numbers as such' appears to be a modest price to pay for all the constructive opportunities such a postulation brings in terms of deriving all the existential consequences of emergence, ascension and physical structuralization. This, I believe, justifies the logical mismatches (and perhaps even some level of absurdity) that such treatment invokes.

To restate, in the brand of metaphysics adopted in the above argument, the ultimate origin of everything lies in the infinite complexity of pure (integer) numbers. Integer numbers form the lowest level of infinite sets, the so-called aleph naught (aleph zero) of Georg Cantor (Dauben (1977), (1979), Rucker (1987), Lavine (1994), Pickover (1995), (2001)). Even this lowest level of infinity (integer numbers) contains in itself an inexhaustible source of complexity—such as the distribution of prime numbers (Dickson 1960, Maier 1981, Ribenboim 1989, Casti 1990, Plichta 1997, Derbyshire 2004) and the digits of pi (Wagon 1985, Preston 1992) or any iterative protocol deduced from it—and all this exerts some kind of a metaphysical pressure for the embodiment of any fragment of the 'infinite resource' of ideal Platonic numerological patterns to transform (or 'convert') these (absolute) patterns into a physical reality.

This immutable all-defining universal 'vacuum' of pure (integer) numbers can perhaps be classified as 'aleph zero panpsychism'. Physical vacuum is relative and may be quite different in the different baby universes of inflationary cosmology and/or the innumerous branches of the Everett's model of ever-breeding quantum universes, while the metaphysical *potentialia* of the Platonic numerological 'vacuum' is absolute and always the same. It does not fluctuate (at any scale) and any of its specific features (e.g. some anomaly in the prime number distribution) is instantly available as a universal morphogenetic (pattern-generating) trigger in any of the innumerable branches of the inflationary cosmos and/or Everett's cosmological foam. This metaphysical *potentialia* does not bother itself with the origin of the physical universe (Gott and Li 1998), or with whether there can be time travel (Gott 2002, Yourgrau, 1999). In Everett-type models the creation of the Universe automatically becomes a non-issue, as with unanswerable questions such as '*who created mathematics?*' or '*why is 17 a prime number?*'. I am leaving the issue of whether higher alephs (uncountable sets) can add new insights to the metaphysical scenario described above as an open question.

## 3.7 It from bit and the Leibnitz principle

*It from Bit symbolizes the idea that every item of the physical world has at bottom an immaterial source and explanation [...] that all things physical are information-theoretic in origin and that this is a participatory universe.*

John Archibald Wheeler (1911–2008)

The great philosopher Gottfried Wilhelm Leibnitz (1646–1716) said that in order to explain the origin of everything from nothing 'there suffices a single principle' (Russell 1989). Models of eternal cosmic inflation pretend to rescind this Leibnitz principle (LP), pointing out that the question of why there is something rather than nothing becomes self-referential. This objection, however, can be objected to. Philosophical Platonism asserts the absolute (pre-)existence of an infinitely rich immutable world of numbers and mathematical structures. This ideal Platonic world (IPW) exists everywhere (but nowhere in particular) and logically precedes space, time, matter or any physics in any conceivable universe.

How the IPW generates physical reality is the central point of the LP. The notions of the Platonic pressure effect (Berezin 1998) and/or 'it from bit' (Wheeler 1990) may be efficient metaphorical tools in envisioning these ideas. One of the key structures of the IPW is an (infinite) hierarchy of Cantor's alephs and, specifically, the sameness of number of integers and rationals (a rational number is the fraction $p/q$ with both $p$ and $q$ being integer numbers, e.g. 2/3, 17/11, 317/137, etc). Such illustrations as 'Cantor's carpet' construct demonstrate the zero Lebesgue measure of rationals on the $x$-axis. An infinite nested hierarchy of alephs resonates with what we often see as the fractal structure of the physical universe.

Asymptotically, at higher cosmic scales the average density of matter seems to tend to zero (a visual analogy can be given by finite-volume zero-mass fractal constructs such as the Menger sponge). Thus, in the spirit of the 'inverse Zeno paradox', one can suggest the possibility of generating $M = 0$ (zero mass) states directly from IPW. In this regard the (meta)physical Platonic pressure of the infinitude of numbers becomes an engine for the self-generation of the physical universe directly out of mathematics.

This, presumably, is the essence of the LP. While physics in other branches of the inflating universe can be (arbitrarily) different from ours, number theory (and the rest of the IPW) is not: it is unique and absolute. For example, pi is expressible as (exactly) the sum of the Leibnitz series that involves only integers (section 2.4), namely it involves an infinitude of all (odd) integers, while in classical Euclidean–Cartesian flat space geometry pi has a clear geometrical meaning as the length of a circle with a unit diameter. Thus, paraphrasing Carl Sagan in his book *Contact* (Sagan 1986), the 'message of pi' may be eventual flatness and an infinity of (embedding) space–time. At the Planck scale the total number of quantum states of 'our' (Big Bang) sub-universe is (well) below $10^{1000}$, which, in terms of aleph naught, is still an infinitely small fraction of all integers. This seems (but just 'seems'—I give no guarantee for any of these models) to favor the steady-state universe (Fred Hoyle) and/or eternal inflation models.

In addressing the eternal question of why there is something rather than nothing, concepts like the zero-point quantum fluctuations (ZPQF) of (whatever) vacuum are often used as alleged vital ingredients in some kind of primordial universal field ('void unfolding' in more explicitly metaphysical systems). However, even ZPQF still implies that some specific 'physics' is involved in universal cosmological dynamics.

Along this line of thought, I propose that we go one step deeper (perhaps to the last step, indeed) and hypothesize that the ultimate origin of everything lies in the infinite complexity of pure (integer) numbers (the aleph zero of Georg Cantor), so this inexhaustible source of complexity (e.g. prime number distribution, the digits of pi, or any iterative protocol deduced from it) exerts some kind of metaphysical pressure for the embodiment of (any of the infinite resource of) ideal Platonic numerological patterns, for the conversion of these (absolute) patterns into a physical reality. This immutable all-defining universal vacuum of pure (integer) numbers (UVPIN) can perhaps be classified as aleph-zero panpsychism.

Likewise, the links between gravity and pure numerology may likely be tractable as well. For example, gravitational fluctuations at the Planck scale might be pertinent for the psi reductions otherwise known as wave function 'collapse' (Penrose 1994, 1996) and self-organization. Here we can put together for comparison the issue of the 'subtleness' of gravitational effects and (the alleged) 'subtle' correlations in seemingly random (normal) digital strings. This is the 'distant Moon' analogy, described metaphorically by Richard Preston (Preston 1992, p 67) to account for the alleged long-scale fluctuations in the digits of pi. Another conceptual umbrella-term analogy is the dichotomy between physical ZPQF and mathematical 'primological' quasi-chaos. In this vista, opposing ZPQF to UVPIN resembles the dichotomy between 'really random' physical noise and pseudo-noise in model systems and 'deterministic chaos', which is structured and reproducible.

Physical vacuum is relative (and it may be quite different in the different baby universes of inflationary cosmology and/or Everett's quantum branches), while the metaphysical *potentialia* of UVPIN is absolute and ever-the-same. It does not fluctuate (at any scale) and any of its specific features (e.g. some anomaly in the prime number distribution) are instantly available as a universal morphogenetic mold in any branch of the inflationary cosmos/Everett's foam of ever-breeding universes.

## 3.8 Mandelbrot sets and the infinite intricacy of the Platonic world

Mandelbrot sets (and their numerous 'relatives', such as Julia sets, etc) are patterns that are produced by chains of iterations of simple-looking algebraic equations. I will not provide details here, as they can be found easily elsewhere. The important point is that these sets have an infinitely intricate fractal structure, in spite of the fact that the equations that generate them are pretty simple. This, among other things, is another illustration of the power of the unfolding of the ideal Platonic world of numbers and forms.

Within the physical context, Mandelbrot sets and other similar models are living examples of the emergence of complexity or self-organization from chaos to order. That is how the infinite complexity and intricacy of the ideal Platonic world translates itself into something that is physical and tangible. While the Mandelbrot set as such is an ideal mathematical construct, it can nonetheless be truncated to real colored pictures, and these are now quite popular for entertainment and various design purposes, from home decoration to fashion.

Likewise, various computer-generated landscapes with mountains, lakes, water-falls, etc can be produced by relatively short and simple algorithms. They look quite realistic and can easily be confused with real photographs of landscapes. Perhaps if we develop the means and instrumentation to 'see' the isotopic structure of material objects on an atom-by-atom basis, we will find some interesting patterns there ('isotopic sculptures' and 'isotopic landscapes'?). I will say more about that in section 4.9 (isotopic superlattices).

The power of information-based unfolding can be seen at many levels of the (so-called) material world. From a small seed grows a huge tree. From two tiny cells of a few microns (not visible to unaided eye) an embryo and then a human being emerges. Einstein was right when he said, 'There are two ways of living life: one is that nothing is a miracle, another is that everything is a miracle'. And we can all, to the best of our abilities and interests, contemplate this universal miracle of informational unfolding from the ideal Platonic world.

## 3.9 Symmetry breaking and the emergence of order

*Science is wonderfully equipped to answer the question 'How?' but it gets terribly confused when you ask the question 'Why?'.*

Erwin Chargaff (1905–2002)

Interesting and somewhat unexpected examples of symmetry breaking and spontaneous ordering can be found even in such a 'simple' area of physics as classical electrostatics. In 1985 this author published a letter in *Nature* (Berezin 1985b) that almost instantly produced a deluge of replies. Imagine a flat and thin circular box (like the box for a round pizza) to which you add several ($N$) point charges. These are small charged balls, and each ball has the same charge (or you can envision classical electrons as point charges). If you add just two charges ($N = 2$), due to electrostatic repulsion they will be at opposite ends of the diameter. This is the result of the energy minimization of the system (a stable configuration has the minimum possible energy). If $N = 3$, the charges inside the circular box will form an equilateral triangle, for $N = 4$ the arrangement will be a square, for $N = 5$ it will be a pentagon, etc. For $N = 11$, the charges will still form a symmetrical 11-gon (all charges on the circumference), but if we add 12 charges to the box ($N = 12$), the minimization of the total electrostatic energy of the system (each charge interacts with every other charge by Coulomb's law) leads to the minimum energy configuration when only 11 charges remain at the circumference, but the 12th charge is expelled to the center of the circle. For $N > 12$, the stable configurations become more complicated, but some of the charges will always 'hang' inside the box (not touching the wall that is the confining circumference).

So, what was the fuss and excitement in *Nature*'s pages about my letter? Here, I have to admit a subtle mistake on my part. First, the title of my letter was somewhat teasing, and some may say even a bit provocative: 'An unexpected result in classical electrostatics'. My (implicit) premise was that charges are confined in a circle. In a two-dimensional world, as in Edwin Abbot's *Flatland* (1884). Almost instantly, *Nature* (20 June and 25 July 1985) published several letters pointing out the fallacy of confusing two- and three-dimensional electrostatics. The authors included such luminaries as Nobel Prize laureate Allan M Cormack and Cambridge cosmologist and futurologist Martin Rees. In the 19 September 1985 issue of *Nature* I published another letter (Berezin 1985c), where I thanked my critics for their insightful comments and pointed out that 'the paradox is, indeed, resolved, if the circle with point charges is treated as a thin disk embedded in three-dimensional space and the central charge is actually at the center of one of the flat surfaces of the disk'.

The discussion of this issue continued for a few more years in the pages of *Nature* and the *American Journal of Physics*, which demonstrated that a healthy fascination with apparent 'paradoxes' may still happen in the science community.

Expanding further on informationally induced patterning, we can see that even the most seemingly random systems have some distinctive morphology and diversity in their structure, dynamics, evolution and responses to external factors. A matter of primary interest is the growing understanding of the possible role of quasi-biological (generic) patterning when the development of a particular morphology exhibits catalytic (dynamically and/or structurally mimicking) features (Berezin 1994b). In other words, it may well be worth looking for some evidence of natural cloning of various 'soft' or plasmatic (dusty-based) structures.

For the purpose of providing a vivid metaphor we can mention Fred's Hoyle 'black cloud' paradigm (Hoyle 1957). Hoyle had earlier suggested the possibility of an informationally evolved interstellar dust cloud generating 'plasmatic offsprings' (see also Berezin 1988). A similar idea of a living planet featured in the novel *Solaris* by Stanislaw Lem (Lem 1981).

Recently, Gregory Sams advanced the idea of a 'living Sun' (the Sun is alive and possesses a consciousness that is far superior to ours), appealing to the ideas of chaos, self-organization and quantum physics. His book *The Sun of gOd* (Sams 2009) produced excitement among a variety of similar-thinking audiences. (The capital 'O' in the last word of the book's title symbolizes the disk of the Sun). According to Sams's ideas, the other stars and entire galaxies are all living beings which (or who?) are communicating with each other, as our Sun is in constant dialogue with her (his?) neighboring sister (brother?) stars, such as Alpha Centauri, Sirius, Vega, etc. That appears pretty hilarious, but the best part of it is that it, indeed, may be a reality of the Universe of which we are a tiny part. At least, there seems to be nothing in physics or logic that rules out such an idea. In the words of the Nobel Prize physicist Murray Gell-Mann, 'Anything which is not prohibited is compulsory' (Comorosan 1974).

The other often-used metaphor for this line of pan-psychistic ('all is alive') ideas is the 'Gaia hypothesis', which pertains to the inner connectedness 'of various, seemingly distinct, species at the most basic informational level (possibly, at a mathematically Platonic, numerological level). An example of such a morphological connectedness is the existence of several distinct species (types) of galaxies. Galaxies show (in a way) quite an analogy with dusty plasma systems. If self-organization is (almost obviously) happening in a world of galaxies (Berezin 1988) one can pose a similar question regarding the world of plasma systems and look for a quasi-evolutionary interpretation of self-organization processes in them too (Berezin 1994b).

Informational quantum connectivity also implies the reformulation of an idea of simultaneity as applied to a dynamical linkage in loosely bound systems. By this we mean a generic time-controlled similarity of underlying types of dynamical morphologies (quantum parallelism). Quantum cloning is usually interpreted as a more subtle (and problematic) effect than classical cloning ('Xeroxing'). Nevertheless, by virtue of an argument of dichotomic negation, it may turn out to be (almost exactly) the other way around.

On the quantum side a mighty pattern-forming factor is quantum non-locality when combined with the fractal structure of morphogenetic structural (or dynamical) attractors (Berezin 1994b). This may excessively offset the recently claimed quantum cloning limitations and lead to pattern proliferation in soft and plasmatic systems, as well as liquid systems. All of this, in a broader context, can be seen as a paradigmatic example of informational holography. Holography generally means that the informational content of the whole is contained in (any of) its parts. This is largely (but not exclusively) a quantum rather than a classical situation. Alternatively, the whole-in-a-part paradigm can be illustrated in a quite straightforward way by genotype–phenotype complementarity in biology.

Closer to the ground, an interesting model (and, perhaps, a test ground) for self-organizational dynamics in crystals can be provided by hopping conductivity. In the hopping mechanism of conductivity, the electrons 'hop' by way of quantum tunneling from one impurity center to another one in the neighborhood of the initial site (Berezin, Golikova and Zaitsev (1973), Berezin (1980b), (1981a), (1981b), (1983a), (1983b), (1983c), (1984), Bottger and Bryksin (1985), Shklovskii and Efros (1984)). Such a process resembles our jumping (hopping) to dry spots over puddles of water after heavy rain. However, for an electron 'choosing' what site to jump to there is 'quantum freedom of choice', since there are several candidate sites in the vicinity to which quantum tunneling can occur.

This 'freedom' in choosing the path of jumps is akin to the 'freedom of choice' in double slit quantum experiments when an electron 'chooses' from which slit to pass. In fact, the analogy will be made with a 'many slits' experiment, since the electrons here have many choices of where to jump to. As a result, the overall path of each particular electron in such a process is a complex zig-zag trajectory with uneven lengths of jumps, random changes of direction, some dead-ends, etc. It should be said, however, that the interpretation of such quantum processes is more subtle and cannot fully rely on classical analogies (Berezin and Nakhmanson 1990, Albert 1992, Deutsch 1997, Penrose, 1994). Thus, in such a process, the collective of zillions of electrons jumping in a complicated dynamic pattern can 'probe' all sorts of informationally loaded arrangements in a somewhat similar way to the firing of random neurons in a human brain.

Talking more about informational self-organization in hopping systems, we can recall that in a partially compensated semiconductor (PCS) the Fermi level is pinned to the donor sub-band. Due to positional randomness and almost isoenergetic hoppings, a donor-spanned electronic subsystem in the PCS forms a fluid-like highly mobile collective state (Shklovskii and Efros 1984). This makes the PCS a playground for pattern formation, self-organization, the emergence of complexity, electronic neural networks, and perhaps even for the origins of life, bio-evolution and consciousness. Through the effects of electronic impact (collisions with sites) and/or Auger ionization of donor sites, a whole PCS may collapse (spinodal decomposition) into micro-blocks that are potentially capable of replication and proto-biological activity (DNA analogue).

Electronic screening effects may also act in an RNA fashion by introducing additional length scale(s) to system. Spontaneous quantum computing on charged/neutral sites becomes a potential generator of informationally loaded microstructures in a manner akin to the 'Carl Sagan effect' (see section 2.4) or the informational self-organization of Borges' *Library of Babel* (section 2.5). Even general relativity effects at the Planck scale (Penrose 1994, 1996) may affect the dynamics through (e.g.) isotopic variations of atomic mass and local density (Berezin 1992a). Thus, a PCS can serve as a toy model (experimental and computational) at the interface of physics and life sciences.

## 3.10 Electrostatic ordering and ionic crystals

Plasma crystallization in particle traps is now a large and growing area of research activity. There was even a Nobel Prize awarded recently (2012: David Wineland and Serge Haroche) for studies on laser-cooled plasma crystals. The potential importance of this work pertains to the use of such artificial crystals for quantum computing, where the preservation of quantum coherence is the key issue. My work on electrostatic ordering, discussed in the previous section, led me to make a modest contribution to plasma crystal studies in 1989 through the paper 'Ordering and patterning of interacting microparticles: confined repulsive particles and isotopic correlations' (Berezin 1989), in which I indicated the analogy with the emergence of correlations in weakly interacting systems. The examples used included isotopic correlations in solid and plasmodic crystals and hypotheses regarding the origin of the law of interplanetary distances (the Titius–Bode law). This law states that the distances of planets (Mercury, Venus, Earth, Mars, etc) from the Sun obey a simple power law as if (if you want to put it metaphorically) the planets 'know' of each other. Such a planetary self-organization has allegedly developed over many millions of years due to the weak tidal effects produced by inter-planetary gravitational interaction (another example of the butterfly effect, where weak interaction can lead to macroscopic consequences).

The philosophy adopted in my papers (Berezin, 1985a, 1986b, 1986d, 1994b, 1997, 2005) on ordering and quasi-crystallization in gas-like plasma systems visualizes Coulombically driven self-organization from the unifying positions of the physics of information. The essence of these effects is that while the Coulomb (electrostatic) interaction is a simple and central force between particles, in the system of many interacting particles it can nonetheless produce some quite intricate patterns of particle arrangement. This is just another illustration of the emergence of complexity in connected systems ('how complexity can arise from simplicity').

In diffuse and highly diluted systems (e.g. low pressure plasmas) the typical size of particles is often several orders of magnitude smaller than the average inter-particle distance. Numerous phenomena related to dust particles, plasmas, gas discharges and similar systems are accompanied by the formation of electrically charged particles. The size, texture, shape and total electrical charge of such particles varies within wide limits. The mutually remote particles interact primarily (and often almost entirely) through long-range Coulomb forces. Particles may also interact through collisions. In addition, a somewhat minor (usually) role can be played by the polarization and van-der-Waals (dispersive) forces. The latter effects introduce some additional (often subtle) nonlinearities into the system which are key for the unfolding of the self-organizational dynamics (Nicolis and Prigogine 1977).

In this vein, I am talking here about some fundamental quantum aspects of self-organization in dynamically random systems (DRSs). A DRS can be defined in a broad sense as a non-periodical and instantaneously random arrangement of interacting entities. This can be understood in quite a broad way as, e.g. atomic or molecular systems, dust particles, stars in galaxies, or living species in populations. A common feature of all DRSs is their dynamical behavior in time; they are not 'fixed' in the same sense as solid-state systems are fixed if the dynamical processes in the latter (e.g. vibrations of crystal lattices) are ignored.

As was said above, it is fundamentally difficult, if not impossible, to define rigorously what is 'random' (Wolfram 1985), we leave it largely to the power of examples to illustrate the term and thus to offset this apparent conceptual impossibility. The examples can be drawn from quite diverse directions, ranging from solid-state systems, such as random interacting impurities or random quantum tunneling in hopping conductivity (Berezin 1980b, 1981a, 1981b, 1983a, 1983b, 1983c, 1984), through to diluted solutions and gaseous or plasma systems (Berezin 1994a). In some of these cases the effects of randomness are juxtaposed with some ordering effects, primarily (but not exclusively) comprised by Coulomb interaction. My purpose here is to present the unifying features of many diversified DRSs and indicate some possible grounds for asserting that there is some inner connectivity in them. This connectivity has an informational background, which in some cases can be related to more fundamental quantum effects such as quantum nonlocalities. This inner connectivity often runs 'hidden' behind apparently random uncorrelated behavior.

Several physically important consequences of Coulomb interaction can be discerned and specified at the present stage of theoretical development. One of the most studied aspects of Coulombic interaction in dusty plasmas is the formation of Coulombically correlated structures. At the same time a somewhat independent (and potentially, perhaps, quite important) line of thought is now emerging. This line can be provisionally called 'informational' as it makes use in a substantial way of the concepts of the physics of information (Landauer 1991). Such concepts as memory effects, learning and adaptive systems, neural networks and self-organization (to name just a few) were previously almost exclusively related to traditionally 'non-physical' domains, such as biology, sociology and their numerous ramifications. This all changed in recent years, mostly (but not exclusively) due to significant progress in the physics of chaos and self-organizing systems, both 'living' and 'non-living'. For the latter, a quasi-biological memory effect has been claimed (Westerlund 1991). Some phenomena in dusty plasmas and other similar spatially extended and/or loosely bound systems can also be fitted into the paradigms of chaology and self-organization.

There are several close links between the concepts of classical chaology, catastrophe theory and a quantum level of consideration. To emphasize these links, we can contrast the concepts of energy and information and extend this comparison to the domain of quantum chaos (Jensen 1992, Gutzwiller 1991, Prange 1992). For the latter, Michael Berry suggested 'quantum chaology' as a conceptually better term (Berry 1989). Informationally induced patterning can then be presented as an effect that inherently has both classical and quantum aspects.

Based on these concepts, we can then look at a range of various effects that contribute to the informationally induced patterning in DRSs, including dusty charged plasmas. I specifically note the informational filtering effect and relate it to energy accumulation mechanisms and soliton formation. *Informational filtering* is interpreted here as a purely energetic mapping and a highly nonlinear energy focusing effect. Our consideration then focuses on a self-induced quantum singularity, which is interpreted as a quantum collapse of the wave function of a DRS with mesoscopic (dust) particles. In addition to this, I include here a few other informational metaphors for self-organizing plasma systems, such as learning (adaptive or neural) networks. Such a context provides a quantum vista for the informational self-organization effects in dusty plasmas and generically similar systems.

Another conceptual framework for such discussions can be achieved through focusing on the (quantum) informational connectivity in loosely bound systems with relation to some ideas regarding how David Bohm's 'holographic paradigm' can possibly be relevant to DRSs in general and gaseous systems like dusty plasmas in particular. This presumes the visualization of Coulombically driven self-organization from the unifying positions of the physics of information. Such an approach in most cases points to the Boltzmann–Shannon definition of entropy in terms of binary informational content (section 2.3). However, here we can look beyond the Shannon equation into a realm of quantum ramifications, especially those related to the paradigms of so-called quantum nonlocalities and quantum automata.

This (contextually and comparatively) invokes the whole range of physical effects and recently developed concepts that can be put in correspondence with the physics of dusty charged self-organizing plasmas and similar systems. These connections could also turn out to be relevant for the study of many interesting and still not well-understood effects and phenomena. Some of these have the reputation of being 'fringe' phenomena, and scientific study of them often faces upfront skepticism. Among them are such occurrences as ball lightning (ter Haar 1989) and 'crop circles vertices', both of which have recently been the subject of a wide range of hypotheses (while crop circles may well be hoaxes, ball lightning almost certainly is not).

The next interesting aspect is to trace a relationship between electrostatic (Coulomb) interaction and the issue of symmetry. Here we enter the area of definitions that are tightly related to our (human) perceptual assessments. The issue of symmetry in this respect is akin to the order-versus-randomness issue. It is, perhaps, quite difficult to define exhaustively (sharply and unequivocally) what 'symmetry' is or what (perfect) 'randomness' is (Wolfram 1985). Fortunately, though, these things (concepts) can be relatively easy subjected to a more-or-less self-consistent quantification grading. Looking for links between Coulomb forces in multiparticle (diffuse) systems and symmetry-forming dynamics, we can observe the following. Symmetry (or 'sameness') is usually understood as the existence of some sort of a movement (transformation) that leaves the system in a certain way 'unchanged'. For example, four points on the surface of a sphere can form a perfect inscribed tetrahedron. All of its vertices are exactly equivalent and their mutual replacement changes nothing. The same can be said about $N = 2, 3, 6, 8, 12$ and $20$

points on the surface of a sphere (the case $N = 1$ is trivial). For $N = 4, 6, 8, 12$ and $20$ the symmetrical placement of points (as vertices) forms five ideal Platonic bodies.

However, for all other integer values of $N$ (e.g. for $N = 5, 11, 28, 37, 500, 1993$, etc (the numbers are taken arbitrarily)) there is no exactly equivalent placement of $N$ points on a spherical surface. If the points are 'physical' and each carries an equal point electrical charge ($Q$), the determination of the minimum energy configuration is quite an intricate problem (Berezin 1986c). The minimum-energy pattern of $N$ equal point charges on a sphere is known for some particular values of $N$, but not for a general $N$. The problem of the exact minimum-energy configuration of $N$ points when some of them carry different values of $Q$ is generally almost intractable using currently existing methods and can be subjected to approximate solutions by randomness-based optimization techniques, e.g. Monte Carlo, travelling salesman algorithms, etc.

For all cases except the above few 'lucky' (Platonic) numbers the minimum-energy arrangement of $N$ point charges can be said to be *quasi*-symmetric (meaning there is some degree of approximation). In the case of a three-dimensional periodical placement (not the surface of a sphere) with translational invariance, this quasi-symmetric positioning is known as quasiperiodicity (Penrose 1989, 1994, 1996). Quasiperiodicity means that the pattern of particles (or other objects) looks 'the same' only FAPP ('for all practical purposes'[2]), that it is only 'at large' in general terms, but not necessarily exactly.

Talking about particles in systems like dusts and molecules (and even atoms) we face the additional issue of 'labelability' (ability to be labeled). This could be roughly defined as the possession of some degree of individuality or distinguishability between objects. Even atomic systems (single atoms) can have some (although quite limited) degree of distinguishability due to their isotopic variations (Berezin 1992b, 1993a). Another (quite nontrivial) source of labelability can be attributed to individuality in a particular (instantaneous) connectivity of a given unit (atom, dust particle, etc), i.e., in how this unit (element) is connected to its other neighbors. Such a 'participation' ('implicit presence') of others in a given element changes its 'personal labels'. This can be, to some extent, compared to mass defect in nuclear physics or in any system of interacting particles when the interaction energy changes (renormalizes) the Einsteinian rest mass of the object in question.

## 3.11 Rene Thom's catastrophe theory and electrostatic phase transitions

Another interesting set of problems arises when we consider external perturbations imposed on a system of confined particles. An example is a system of a few charged particles inside a sphere with the imposition of an external electric field (Berezin 1986b 1986d). An analysis demonstrates that there will be a sharp (discontinuous) configurational rearrangement when some critical value of the imposed electric field is reached. This can be called an electrostatic phase transition. The model for

---

[2] The 'FAPP' abbreviation comes from John Bell.

the adequate and vivid description of such an effect is Rene Thom's catastrophe theory (Berezin 1991).

Catastrophe theory is generally seen as being somewhat to the side of mainstream physics and mathematics for its utter simplicity in describing sharp (singular) transitions in a variety of systems. However, the realm it covers is surprisingly large—the 'real' physics of phase transitions, the stability and collapse of engineering structures, the capsizing of ships, psychology and criminality (riots in prisons, etc), sports and the stock market, the fight-or-flight behavior of animals, falling in love and the arts (e.g. Zeeman 1977, Saunders 1980).

My own (somewhat unexpected) contribution to the applications of catastrophe theory includes the dynamics of the corrosion process, such as the set-up of pitting corrosion (De Sa and Berezin, 1989; Berezin, 1993a, 1993b), and modeling of the phase transitions and structural transformations in systems of ionic crystals and charged plasmas (Berezin 1986b, 1986d, 1989, 1991, 1994b, 1997).

Electrostatic phase transitions refer to sudden ('singular') structural rearrangements in the systems of the confined charge particles. According to the theorem proven by Samuel Earnshaw in 1842, the system of electrostatically interacting particles cannot be in a stable stationary equilibrium configuration solely through the electrostatic interaction of the charges. The reason for this is easy to see from the minimum energy principle. In a simple system of two free point charges, if they have the same sign (say, positive), the charges will repel each other and fly from each other to infinity to minimize their potential energy to zero. Conversely, if the point charges are of opposite sign they will Coulombically attract each other and fall on each other to zero distance, and their potential energy will go to minus infinity. Therefore, the stability of system of charges requires that some additional forces and/or interactions of a non-electrostatic nature be acting to assure the stable static configuration of the charged particles. A confining surface is the most obvious example of such an extra force.

Let us provisionally classify all the internal interactions in dilute gaseous systems and some other dynamically random systems as consisting primarily of two components: collisions and Coulomb interactions between charges. In this context a DRS is defined in a broad sense as a non-periodical and instantaneously random arrangement of interacting entities (Berezin 1994b). In addition to the collisional and Coulomb interactions, several 'external' interactions can be indicated that may act on DRSs: (1) the effects of gravity, (2) the effects of externally imposed (solely or in any combination) electric and magnetic fields (static, quasi-static or oscillating), including excitation by light, and (3) the effects of macroscopic gradients of external parameters, such as, e.g. pressure, the partial (osmotic) pressures of components, etc.

All these factors jointly define the dynamical evolution of a system. At the phenomenological level, (any) dynamics can be described in a number of ways. These can be alternatives or complementary to one another. Let us consider, for example, the typical microscopic processes that can occur in a (non-stationary) dust particle system. In the presence of electromagnetic radiation, particles can scatter through, e.g. the well-known mechanism of Rayleigh scattering. This process is similar in a number of ways to the hopping of electromagnetic radiation in a system of randomly located stochastic sites or radiation transfer in DRSs (e.g. in solar plasma).

Auxiliary effects to such inter-bouncing of electromagnetic waves can include a range of phenomena observed for non-stationary dusty plasmas. Such a (well known) phenomenon as luminescence can be mentioned as an example. Because of the various relative velocities both Stokes and anti-Stokes broadening can participate in such a process[3].

The interconversion of electromagnetic energy in dusty ('soft') systems has some level of inelasticity. This, in turn, reveals a dynamics that is similar to the phonon-assisted-hopping in solids with randomly located impurity centers (Bottger and Bryksin 1985). In a quantum regime the disorder in solids is known to produce quantum localization (Anderson localization), which can be further stabilized by polarization effects (Berezin 1983a). In DRS this may translate to the quasi-localization (quasi-freezing) of local structural arrangements which (due to quantum effects) may exhibit much greater survival times (Prentiss 1993) than the lifetimes that can be expected from the completely classical (non-quantum) kinematics.

Archetypically, the above-sketched kinetics can be fitted into a number of algorithmic formalization schemes. Some examples are 'self-cellularization' (models based on the paradigm of cellular automata) or some aspects of informational dynamics in neural network site models (like Hopfield's spin glass neural network model). In the presence of imposed fields, e.g. constant or almost constant electric fields, the dynamical problem acquires dimensions of self-organization and emergence. It is coupled with the lifetime spectrum of (instantaneous) interparticle local configurations. The latter may exhibit the fractal structure of self-similarity at different spatial and/or temporal scales of dynamics.

Electrostatic phase transitions in $N$-particle systems were discussed earlier (Berezin 1986a, 1986b, 1986d). For the case of a small number of confined charges ($N = 4$) it was shown that at some (critical) values of electrostatic fields sudden ('catastrophic') structural transformations may occur. This was later presented as a case that falls into a typical scenario of Rene Thom's catastrophe theory (Berezin 1991). It is also pertinent at this point to recall that even purely classical systems interacting through the Newtonian potential can exhibit non-collisional singularities in their dynamics (Stewart 1992). The latter means spontaneous energy accumulation on a particular particle in a system occurring in the same way as energy focusing in solitonic modes in liquid and gaseous systems ('soliton' means that the excitation energy is concentrated in a single wave or a vibration). The phenomenology of solitonic energy accumulation in DRS has common ground with shock wave generation and

---

[3] As a reminder: when a system (molecule or a crystal) absorbs a photon, it gains energy and enters into an excited state. When the emitted photon (luminescence) has less energy than the absorbed photon, this energy difference is the Stokes shift. Conversely, if the emitted photon has more energy than the initially absorbed photon, the energy difference is called an anti-Stokes shift; this extra energy comes from the dissipation of thermal vibrations (phonons) in a crystal lattice.

the latter is known to stimulate the specific type of luminescence, sonoluminescence (Johnston 1991).

For dusty DRSs these phenomena can be further enriched by the formation of mesoscopic clusters of dust particles, loosely correlated with Coulomb forces. This opens up the possibility of specific ('magic') configurations of clusters with enhanced stability and greater lifetimes (Jiang *et al* 1993). All these effects may contribute to the pattern-selection mechanisms and self-organizational dynamics of DRSs.

To conclude this section, we observe that electrostatic phase transitions in the DRSs of charged particles are relatable to hopping processes in systems like doped semiconductors (Bottger and Bryksin 1985, Shklovskii and Efros 1984). For the latter the dynamical occurrence of correlated hopping complexes may be fundamental. Additionally, the presence of magnetic fields may lead to the manifestation of macroscopic quantum phenomena like the Aharonov–Bohm effect for the magnetic (Silverman 1993) as well as electrostatic fields (Matteucci and Pozzi 1985).

## 3.12 The problem of '$N$ dictators'

*With the possible exception of the equator, everything begins somewhere.*
C S Lewis (1898–1963)

Many people are fascinated by fully symmetrical regular Platonic polyhedrons, otherwise known as ideal Platonic bodies. There are five of them: the tetrahedron, octahedron, hexahedron (cube), icosahedron and dodecahedron. The number of vertices for them is $N = 4$, 6, 8, 12 and 20, respectively. These are the only symmetries that can be inscribed inside a spherical surface such that all vertices are placed in exactly equivalent positions. That means that if we take $N = 4, 6, 8, 12$, or 20 points, they can be placed fully symmetrically on a sphere, while for all other values of $N$ this cannot be done and some points will have more 'favorable' or 'unfavorable' positions than the others (in fact two or three points can also be put on a sphere symmetrically, but these numbers do not form any objects of volume).

Why it is this way, or why 'Nature' has made it so that only these numbers can be put on a sphere symmetrically, is perhaps a question that belongs to the same metaphysical category as the question of why pi is not exactly three (see section 2.4). Anyway, these ideal Platonic bodies have produced many artifacts in art and jewelry, such as quartz crystal sets, earrings, dice, etc; I have even seen a municipal waste dumpster in the form of an icosahedron. This, together with many other things that are discussed in this book, shows our fascination with perfect symmetries.

So, what is the 'problem of $N$ dictators'? It was formulated by the Hungarian mathematician Fejes Toth in 1949 (Berezin 1986d and references therein) in the following way. Imagine a spherical planet (without oceans) that is governed by $N$ mutually inimical dictators. How should they locate their residences in order to be as far as possible from one another? Or, equivalently, how can $N$ fuel depots be arranged on such a planet so that an accidental explosion at one of them will least endanger the rest (Berezin 1986d)?

An interesting extension of this problem is when particles interact via the non-Coulombic law (force is $1/r^n$, $n > 1$). In this case, as analysis shows (Berezin 1986d), for some combination of $N$ (the number of particles) and $n$ (the parameter in non-Coulombic law) there will be an 'ejection' of the particles from the surface and we will have some particles 'hanging free' inside in the volume.

# References

Albert D Z 1992 *Quantum Mechanics and Experience* (Cambridge, MA: Harvard University Press)

Berezin A A, Golikova O A and Zaitsev V K 1973 Nature of hopping conduction in (beta)-rhombohedral boron *Fizika Tverdogo Tela* **15** 1856–9 (in Russian)

Berezin A A, Golikova O A and Zaitsev V K 1973 Nature of hopping conduction in (beta)-rhombohedral boron *Sov. Phys.—Solid State* **15** 1237–9 (Engl. transl.)

Berezin A A 1980a Radiative tunnel transitions in some negatively charged colour center systems in alkali halide crystals *J. Phys. C: Solid State Phys.* **13** L103–6

Berezin A A 1980b Radiative tunnel transitions in the hopping conduction in doped semi-conductors in a strong electric field *J. Phys. C: Solid State Phys.* **13** L947–9

Berezin A A 1981a On the hopping conduction mechanism of amorphous semiconductors in a strong electric field *Phys. Lett.* **86A** 480–2

Berezin A A 1981b On the theory of the hopping conduction in beta-rhombohedral boron in a strong electric field *J. Less-Common Met.* **82** 143–8 (To the best of my knowledge, this was the first published paper to propose that the isotopic disorder between stable isotopes of boron, $^{10}$B and $^{11}$B may cause electron localization in a complex crystal structure of beta-boron)

Berezin A A 1983a Spontaneous tunnel transitions induced by redistribution of trapped electrons over impurity centers *Z. Nat.forsch.* A **38** 959–62

Berezin A A 1983b Resonance energy transfer in activationless hopping conductivity *Phys. Rev. Lett.* **50** 1520–3

Berezin A A 1983c Energy transfer processes in non-ohmic hopping conductivity in a strong electric field *Phys. Lett.* A **97** 105–7

Berezin A A 1984 Double tunnel jumps in activationless hopping conductivity *J. Phys. C: Solid State Phys* **17** L393–7

Berezin A A 1985a Spontaneous symmetry breaking in classical systems *Am. J. Phys.* **53** 1036–7

Berezin A A 1985b An unexpected result in classical electrostatics *Nature* **315** 104

Berezin A A 1985c The distribution of charges in classical electrostatics *Nature* **317** 208

Berezin A A 1986a Electrostatic stability and instability of $N$ equal charges in a circle *Chem. Phys. Lett.* **123** 62–4

Berezin A A 1986b A discontinuous symmetry change in electrostatic systems *J. Electrost.* **18** 193–7

Berezin A A 1986c Asymptotics of the maximum number of repulsive particles on a spherical surface *J. Math. Phys.* **27** 1533–5

Berezin A A 1986d Simple electrostatic model of the structural phase transition *Am. J. Phys.* **54** 403–5

Berezin A A 1987 Super super large numbers *J. Recreat. Math.* **19** 142–3

Berezin A A 1988 Bubble distribution of galaxies: evidence for bio-evolution? *Specul. Sci. Technol.* **11** 197

Berezin A A 1989 Ordering and patterning of interacting microparticles: confined repulsive particles and isotope correlations *Proc. Workshop Crystalline Ion Beams (Darmstadt, Federal Republic of Germany, April 1989)* ed Hasse R W, Hofmann I and Liesen D pp 255–8

Berezin A A and Nakhmanson R S 1990 Quantum mechanical indeterminism as a possible manifestation of microparticle intelligence *Phys. Essays* **3** 331–9

Berezin A A 1991 Application of catastrophe theory to phase transitions of trapped particles *Phys. Scr.* **43** 111–5

Berezin A A 1992a Correlated isotopic tunneling as a possible model for consciousness *J. Theor. Biol.* **154** 415–20

Berezin A A 1992b Isotopicity: implications and applications *Interdiscip. Sci. Rev.* **17** 74–0

Berezin A A 1993a Isotopic effects and corrosion of materials *Mater. Phys. Chem.* **34** 91–100

Berezin A A 1993b Quantum-mechanical aspects of corrosion dynamics *Trends in Corrosion Research* vol 1 (Kaithamukku: Research Trends, Council of Scientific Integration) pp 67–89

Berezin A A 1994a Ultra high dilution effect and isotopic self-organisation *Ultra High Dilution. Physiology and Physics* ed P C Endler and J Schulte (Dordrecht: Kluwer) pp 137–69

Berezin A A 1994b Quantum aspects of self-organization in dynamically random systems *Dusty and Dirty Plasmas, Noise, and Chaos in Space and in the Laboratory* ed H Kikuchi (New York: Plenum) pp 225–40

Berezin A A 1997 Coulomb correlation effects and pattern formation in electrostatic systems with multistabilities and breakdown *J. Electrost.* **40/41** 79–84

Berezin A A 1998 Meaning as self-organization of ultimate reality *Ultimate Real. Meaning* **21** 122–34

Berezin A A 2004 Ideas of multidimensional time, parallel universes and eternity in physics and metaphysics *Ultimate Real. Meaning* **27** 288–14

Berezin A A 2005 Quantum effects in electrostatic precipitation of aerosol and dust particles *Air Pollution XIII Conf., WIT Trans. Ecol. Environ.* vol 82 ed C A Brebbia pp 509–518 (Boston, MA: WIT)

Berezin A A 2015 *Isotopicity Paradigm: Isotopic Randomness in the Digital Universe* (Cambridge: Cambridge International Science)

Berry M V 1989 Quantum chaology, not quantum chaos *Phys. Scr.* **40** 335–6

Bloch W G 2008 *The Unimaginable Mathematics of Borges' Library of Babel* (Oxford: Oxford University Press)

Borges J L 1998 *Collected Fictions* (New York: Viking)

Bottger H and Bryksin V V 1985 *Hopping Conduction in Solids* (Berlin: VCH)

Casti J L 1990 *Searching for Certainty* (New York: William Morrow)

Comorosan S 1974 The measurement problem in biology *Int. J. Quantum Chem. Biol. Symp.* **1** 221–8

Dauben J W 1977 Georg Cantor and Pope Leo XIII: mathematics, theology, and the infinite *J. Hist. Ideas* **38** 85–108

Dauben J W 1979 *Georg Cantor: His Mathematics and Philosophy of Infinite* (Princeton, NJ: Princeton University Press)

De Sa M S and Berezin A A 1989 The application of catastrophe theory to corrosion problems *Corros. Sci.* **29** 1141–8

Derbyshire J 2004 *Prime Obsession: Bernhard Riemann and the Greatest Unsolved Problem in Mathematics* (New York: Plume)

Deutsch D 1997 *The Fabric of Reality* (London: Allen Lane)

Dickson L E 1960 *Modern Elementary theory of Numbers* (Chicago, IL: University of Chicago Press)

Dyson F J 1979 Time without end: the physics and biology in an open universe *Rev. Mod. Phys.* **51** 447–60

Giordano P 2011 *The Solitude of Prime Numbers* (Toronto: Penguin)

Gott J R and Li X L 1998 Can the Universe create itself? *Phys. Rev.* D **58** 023501

Gott J R 2002 *Time Travel in Einstein's Universe: The Physical Possibilities of Travel Through Time* (New York: Houghton Mifflin)

Gutzwiller M C 1991 *Chaos in Classical and Quantum Mechanics* 2nd edn (Berlin: Springer)

Hersh R 1995 Fresh breezes in the philosophy of mathematics *Am. Math. Mon.* **102** 589–94

Jensen R V 1992 Quantum chaos *Nature* **335** 311

Jiang T, Kim C and Northby J A 1993 Electron attachment to helium microdroplets: creation induced magic? *Phys. Rev. Lett.* **71** 700–3

Johnston A C 1991 Light from seismic waves *Nature* **354** 361

Knuth D E 1976 Mathematics and computer science: coping with finiteness *Science* **194** 1235–42

Landauer R 1991 Information is physical *Phys. Today* **44** 23

Lavine S 1994 *Understanding the Infinite* (Cambridge, MA: Harvard University Press)

Lem S 1981 *Solaris* (New York: Penguin)

Lewis D 1986 *On the Plurality of Worlds* (Oxford: Blackwell)

Lloyd S 2006 *Programming the Universe: A Quantum Computer Scientist Takes on the Cosmos* (New York: Knopf)

Maier H 1981 Chains of large gaps between consecutive primes *Adv. Math.* **39** 257–69

Martinez A A 2012 *The Cult of Pythagoras: Mathematics and Myths* (Pittsburg, PA: University of Pittsburg Press)

Matteucci G and Pozzi G 1985 New diffraction experiment on the electrostatic Aharonov–Bohm effect *Phys. Rev. Lett.* **54** 2469–72

Nicolis G and Prigogine I 1977 *Self-Organization in Nonequilibrium Systems: From Dissipative Structures to Order Through Fluctuations* (New York: Wiley)

Penrose R 1989 Tilings and quasi-crystals; a non-local growth problem? *Introduction to the Mathematics of Quasicristals* ed M V Jaric (New York: Academic) pp 53–79

Penrose R 1994 *Shadows of the Mind* (Oxford: Oxford University Press)

Penrose R 1996 On gravity's role in quantum state reduction *Gen. Relativ. Gravit.* **28** 581–600

Pickover C A 1995 *Keys to Infinity* (New York: Wiley)

Pickover C A 2001 *Wonders of Numbers: Adventures in Mathematics, Mind, and Meaning* (New York: Oxford University Press)

Pickover C A 2007 *A Beginner's Guide to Immortality* (New York: Thunder's Mouth Press)

Plichta P 1997 *God's Secret Formula: Deciphering the Riddle of the Universe and the Prime Number Code* (Rockport, MA: Element)

Prange R E 1992 Topics in quantum chaos *Chaos and Quantum Chaos* ed W D Heiss (Berlin: Springer) pp 225–71

Prentiss M G 1993 Bound by light *Science* **260** 1078

Preston R 1992 The mountains of pi *The New Yorker* (2 March 1992) pp 39–67

Ribenboim P 1989 *The Book of Prime Number Records* 2nd edn (New York: Springer)

Rucker R 1987 *Mind Tools (The Five Levels of Mathematical Reality)* (Boston, MA: Houghton Mifflin)

Russell B 1989 *A History of Western Philosophy* (London: Unwin) (and many other editions)

Sagan C 1986 *Contact* (New York: Pocket)

Sams G 2009 *Sun of gOd* (San Francisco, CA: Weiser)

Sangalli A 2006 *Pythagoras' Revenge: A Mathematical Mystery* (Princeton, MA: Princeton University Press)

Saunders P T 1980 *Catastrophe Theory* (Cambridge: Cambridge University Press)

Shklovskii B I and Efros A L 1984 *Electronic Properties of Doped Semiconductors* (Berlin: Springer)

Silverman M P 1993 More than one mystery: quantum interference with correlated charged particles and magnetic fields *Am. J. Phys.* **61** 514–23

Stewart I 1992 Cosmic tennis blasts particles to infinity *New Scientist* (3 October 1992) **136** p 14

Tegmark M 2014 *Our Mathematical Universe: My Quest for the Ultimate Nature of Reality* (New York: Knopf)

ter Haar D 1989 An electrostatic-chemical model of ball lightning *Phys. Scr.* **39** 735

Tipler F J 1994 *The Physics of Immortality: Modern Cosmology, God and Resurrection of the Dead* (New York: Doubleday)

Tsytovich V N, Morfill G E, Fortov V E, Gusein-Zade N G, Klumov B A and Vladimirov S V 2007 From plasma crystals and helical structures towards inorganic living matter *New J. Phys.* **9** 263

Wagon S 1985 Is pi normal? *Math. Intell.* **7** 65–7

Westerlund S 1991 Dead matter has memory! *Phys. Scr.* **43** 174

Wheeler J A 1990 Information, physics, quantum: the search for links *Complexity, Entropy, and the Physics of Information* ed W H Zurek (Redwood City, CA: Addison-Wesley) pp 3–27

Wilczek F 1999 Getting its from bits *Nature* **397** 303–6

Wolfram S 1985 Origins of randomness in physical systems *Phys. Rev. Lett.* **55** 449

Yourgrau P 1999 *Gödel Meets Einstein: Time Travel in the Gödel Universe* (Chicago, IL: Open Court)

Zeeman E C 1977 *Catastrophe Theory—Selected Papers, 1982–1977* (Reading, MA: Addison-Wesley)

**IOP** Publishing

Digital Informatics and Isotopic Biology
Self-organization and isotopically diverse systems in physics, biology and technology
**Alexander Berezin**

# Chapter 4

## Isotopicity in physics and engineering

*The measure of greatness in a scientific idea is the extent to which it stimulates thought and opens up new lines of research.*

Paul Dirac (1902–1984)

This chapter gives a broad and interdisciplinary overview of isotopicity and isotopic effects in physics, material sciences, informatics and nanotechnology. The biological effects of isotopicity are discussed in the next chapter (Chapter 5). To restate what was said above regarding the term isotopicity, by this word (I believe, I am the author of this term, although I cannot claim this with certainty) I mean a unifying idea in which the isotopic diversity of chemical elements and the effects coming from this are considered as a single (holistic) phenomenon of Nature, rather than as a toy box of scattered facts. In other words, my intention is to make the reader familiar with what isotopes are and why isotopic diversity (isotopicity) makes a kind of a special world of its own by offering a bird's eye view of the subject. We all are mixtures of a variety of stable (and sometimes radioactive) isotopes and if isotopic diversity (and hence, isotopic randomness) disappeared (that is, if there were only one stable isotope for each chemical element), our world would be entirely different (or, perhaps, there would not be any world at all). After a brief introduction to isotopes, I will review a few ideas related to isotopic engineering. I do not claim to be the sole creator of isotopic engineering (see, e.g. Haller 1995, 2002), yet the very list of my publications in this area gives me, I believe, a platform to talk with some authority about this area of engineering and (emerging) technology.

## 4.1 Stable and radioactive isotopes

*Be less curious about people and more curious about ideas.*

Marie Curie (1867–1934)

The notion of isotopes is generally a well-known fact in many seemingly disjointed areas of science and technology, such as the nuclear industry, isotope labeling in biology and medicine (Berezin 1987c), isotope geology (Berezin 1988d), etc. However, there have been relatively few attempts to look at isotopic diversity as a single phenomenon of Nature. One can somewhat provisionally call this phenomenon 'isotopicity'. Within it there are a variety of notions, such as isotopic diversity, isotopic randomness, isotopic ordering, isotopic engineering, etc, as discussed in this book.

Most of this material is taken from my numerous papers, previously published in a broad variety of journals (Berezin (1981a), (1981b), (1982b), (1984d), (1984e), (1984f), (1984g), (1984h), (1986a), (1986b), (1987a), (1987b), (1987c), (1987d), (1988a), (1988b), (1988c), (1988d), (1988e), (1989a), (1989b), (1990), (1991), (1992a), (1992b), (1993a), (1993b), (1994a), (1994b), (1994c), (2004a), (2004b), (2011), Goldman and Berezin (1995), Berezin and Ibrahim (1988), (1991), Berezin *et al* (1988), Pui and Berezin (2001)). Some of my work on isotopic effects was also done with experimental groups (e.g. Chen, Chang, Berezin, Ono and Teii 1991, Berezin, Zaitsev, Kazanin and Tkalenko 1972, Arseneva-Geil, Berezin and Melnikova 1976).

The educated general public usually associates the term 'isotopes' with radioactivity, the nuclear industry, radioactive contamination and the use of radioactive isotopes in medicine. At the same time, it is well known (but not always properly appreciated) that the majority of chemical elements are mixtures of two or more stable isotopes. With the exception of phosphorus, all of the elements that are essential for the functioning of biological systems are polyisotopic elements. For instance, carbon and nitrogen each have two stable isotopes ($^{12}C$ and $^{13}C$ or $^{14}N$ and $^{15}N$), while oxygen has three ($^{16}O$, $^{17}O$ and $^{18}O$), etc. Numbers like 12, 13, 14, etc before the traditional symbol of a chemical element indicate the atomic weight of a particular atom. The latter is the total number of protons and neutrons in this atom. For example, $^{12}C$ (carbon-12) has six protons and six neutrons in its nucleus, while $^{13}C$ has six protons and seven neutrons. There is also carbon-14 ($^{14}C$), which is a radioactive isotope of carbon with a half-life of about 5000 years. It has six protons and eight neutrons and is widely use for the carbon dating of old objects and archaeological artifacts.

There are currently over 115 known chemical elements (both natural and synthesized), of which 80 elements have stable isotopes, while 21 of them have only one stable isotope. Thus, about 2/3 of stable elements occur naturally on Earth in multiple stable isotopes (figure 4.1). The 'champion' of isotopicity, tin (Sn) has ten known stable isotopes. Altogether, there are 283 stable isotopes for 80 stable elements (about $283/80 = 3.5$ on average per element).

All elements, without exception, have numerous (often several dozen) radioactive isotopes, most of which have short lifetimes (seconds, milliseconds, etc), and the total number of known isotopes is about 3000 (about 2700 of these nucleotides are radioactive).

As is commonly known, isotopes are atoms that have the same atomic number ($Z$) but different numbers of neutrons ($N$) in their nuclei. Therefore, their atomic weights ($A$) are distinctly different—even for the heaviest stable elements with $A >$

| 0 | Pm | Tc |    |    |    |    |    |    |    |    |    |    |    |    |    |    |    |    |    |    |
|---|----|----|----|----|----|----|----|----|----|----|----|----|----|----|----|----|----|----|----|----|
| 1 | Be | F  | Na | Al | P  | Sc | Mn | Co | As | Y  | Nb | Rh | I  | Cs | Pr | Tb | Ho | Tm | Au | Th | Bi |
| 2 | H  | He | Li | B  | C  | N  | Cl | V  | Cu | Ga | Br | Rb | Ag | In | Sb | La | Eu | Lu | Ta | Re | Ir | Tl |
| 3 | O  | Ne | Mg | Si | Ar | K  |    |    |    |    |    |    |    |    |    |    |    |    |    |    |    |
| 4 | S  | Cr | Fe | Sr | Ce | Pb |    |    |    |    |    |    |    |    |    |    |    |    |    |    |    |
| 5 | Ti | Ni | Zn | Ge | Zr | W  |    |    |    |    |    |    |    |    |    |    |    |    |    |    |    |
| 6 | Ca | Se | Kr | Pd | Er | Hf | Pt |    |    |    |    |    |    |    |    |    |    |    |    |    |    |
| 7 | Mo | Ru | Ba | Nd | Sm | Gd | Dy | Yb | Os | Hg |    |    |    |    |    |    |    |    |    |    |    |
| 8 | Cd | Te |    |    |    |    |    |    |    |    |    |    |    |    |    |    |    |    |    |    |    |
| 9 | Xe |    |    |    |    |    |    |    |    |    |    |    |    |    |    |    |    |    |    |    |    |    |
| 10| Sn |    |    |    |    |    |    |    |    |    |    |    |    |    |    |    |    |    |    |    |    |    |

**Figure 4.1.** Chart of the chemical elements from hydrogen ($Z = 1$) to bismuth ($Z = 83$) arranged according to the number of their stable isotopes. The point to notice is that some of the included isotopes are, in fact, slightly radioactive. For example, selenium 82Se has a lifetime of $10^{20}$ years against double beta-decay. Isotopes with lifetimes of the order of the presumed age of the ('Big Bang') universe (about 14 billion years) are commonly counted as stable. Thus, thorium ($Z = 90$) is also added in the first row as having one stable isotope because its main isotope $^{232}$Th has a half-decay time of 14 billion years. Note that the two elements within the stable range of $Z$ have no stable isotopes at all: technetium, Tc ($Z = 43$) and promethium, Pm ($Z = 61$). A somewhat curious fact that both 43 and 61 are prime numbers—whether it is a mere coincidence or if there is some deeper underlying ('numerological') reasons why both these $Z$'s are prime numbers is anybody's guess at this point.

200 the addition of just one extra neutron adds about 0.5% (1/200) of a mass—clearly a macroscopic increase by all quantitative standards. Therefore, for a variety of mass-related processes the difference between isotopes of the same element should not be perceived to be a subtle and insignificant 'higher order' correction, but should rather be seen as a zero order and easily detectable effect (e.g. Hoffman and Scherz 1990, Plekhanov 2004).

While there are 81 stable elements in the periodic table with $Z$ ranging from $Z = 1$ to $Z = 83$ (Bi), it is a somewhat curious fact that two elements inside the stable range of $Z$ do not have stable isotopes at all, namely, technetium ($Z = 43$) and promethium ($Z = 61$). To add to this curiosity, one may observe that both 43 and 61 are prime numbers. This author is not aware of any explanation or interpretation of this curiosity or coincidence.

When we say that 21 elements (about 1/4 of the total) have only one stable isotope (e.g. F, Al, Na, P, Mn, Au, etc), we need to make a small correction. In fact, a few elements that we normally count as having stable isotopes are slightly radioactive with extremely long life-times. For example, bismuth (Bi) is normally considered to be the last stable element (Bi has $Z = 83$ and from $Z = 84$ all elements are radioactive). However, it has recently been discovered that the only naturally occurring isotope of bismuth, $^{209}$Bi (83 protons and 126 neutrons), is, in fact, slightly radioactive. It decays through alpha-decay, with a lifetime $2 \times (10^{19})$ years.

This lifetime is about a billion times longer than the presumed age of the (Big Bang) universe (the latter is now estimated at some 15 billion years).

However, a simple estimate using Avogadro's number shows that in one kilogram of bismuth one such decay occurs every three or four minutes. To put it in perspective, there are naturally occurring radioactive isotopes of carbon ($^{14}C$) and potassium ($^{40}K$) in the environment, including all living matter. These isotopes produce a few thousand radioactive decays every second in a human body throughout its entire life and this is the natural background radiation we all have to live with. But not to worry—this has some positive aspects to it, like radioactive hormesis (strengthening of the immune system) or contributions to intuition and creativity (see sections 5.4, 5.8 and 5.14).

These basic facts about isotopes are, of course, generally well known. There have been numerous studies of the effects of isotopes in physics, chemistry and biology. Such disciplines as isotopic geology have emerged on the basis of the isotopic diversity of certain chemical elements. Isotope enrichment and isotope labeling techniques are widely used, not to mention the role of radioactive isotopes in medicine. Nevertheless, despite all this attention to the isotopes of individual elements, there have been very few, if any, attempts to look at the whole notion of isotopicity as a special facet of Nature, as a unique and potentially complex phenomenon in its own right.

This latter view is adopted in the present book. In my papers (and in this book) I offer the view that isotopicity in Nature is more than just a chemical curiosity. My aim is to move this holistic view of isotopicity one step further to see what possible playground it may present for Nature's 'quest for patterns'. By this, I mean that I shall explore what possible opportunities there may be in terms of the pattern generation that may be hidden in the commonly known fact of the isotopic diversity of chemical elements. In other words, I want to look at the possible place of isotopicity in the hierarchy of the structural levels of organization of matter, life and consciousness.

## 4.2 Isotopicity meme

*As for memes, the word 'meme' is a cliché, which is to say it's already a meme. The man who invented it was Richard Dawkins, who was, not coincidentally, an evolutionary biologist. And he invented it as an analog for the gene [...] Memes can be visual. Our image of George Washington is a meme. We don't actually have any idea what George Washington looked like. There are so many different portraits of him, and they're all different. But we have an image in our head, and that image is propagated from one place to another, from one person to another.*
James Gleick (b 1954), author of *Chaos: Making a New Science*

In the view of this author, the idea of isotopicity has the quality and nature of a meme. The concept of the meme was introduced by Richard Dawkins in 1976 in his book *The Selfish Gene* (Dawkins 1976) as an efficient working metaphor with which to describe the propagation of almost anything in human society. While the origin of this concept lies in evolutionary biology as a presumed vehicle for the propagation of (generally useful) mutations, its use can be extended to many other areas, including fashion, home decoration, dominant architectural styles and even popular or 'cool' conversational idioms. From the physics standpoint, the dynamics of meme propagation is similar to that of the propagation of light waves according to Huygens' principle (each point of the light front becomes a source of the secondary emission). Our conceptual framework calls for the connection of these ideas to the notions of mathematical Platonism and Jungian archetypes.

Although in a wider sense the term isotopicity can refer to both stable and radioactive isotopes, in the present book we are focusing primarily on stable isotopic diversity. Two complementary sides of isotopic diversity are of primary interest at the micro and nano levels. These are isotopic randomness and isotopic correlations. If and how the isotopicity meme takes off, and whether it will have a life of its own, is not for me to predict. As the history of other memes shows, they do not always develop and evolve in ways predicted by their originators. And, as I have already pointed out above, I do not claim to be the originator or 'godfather' of this term. Perhaps somebody else used it before me. And I do not so much care for the *term* (isotopicity or whatnot), as for the *concept* behind it, which, as I am trying to argue in this book, has a unifying and heuristic power.

## 4.3 Milestones of isotopes

*Suppose you make a hole in an ordinary evacuated electric light bulb and allow the air molecules to pass in at the rate of 1 000 000 a second, the bulb will become full of air in approximately 100 000 000 years.*

Francis William Aston (1877–1945)

In terms of the history of science the existence of stable isotopes is a relatively recent discovery. Joseph Thomson—the same 'J J Thomson' (1856–1940) who discovered the electron—separated two stable isotopes of neon, $^{20}$Ne and $^{22}$Ne, in 1912 and the very term 'isotopes' was coined in 1913 by another great scientist, Frederick Soddy (1877–1956). Since 'iso' means 'same' and 'top' means 'place', the word isotope is an excellent choice to indicate two or more entities (atoms) which occupy the same place in the periodic table of elements. In the following years, many more stable isotopes were discovered by Thomson's former associate Francis W Aston (1877–1945) and other scientists.

It is worth observing that the original experimental evidence for isotopic mass variations was obtained for heavy and rare radioactive elements (thorium and uranium) and not for the isotopes of the more common lighter chemical elements. That may appear a bit curious, since for lighter atoms the relative isotopic mass differences are much greater than for heavier elements. For example, for the $^{6}$Li and $^{7}$Li isotopes the mass difference is about 15% (not to mention H and D, for which this mass difference is 100%).

Deuterium—which is, perhaps, the most technologically important isotope—was only discovered by Harold Urey (1893–1981) in 1932. However, it is also interesting to note that long before the actual discovery of isotopes by Soddy, Sir William Crookes (1832–1919) had speculated about their possible existence when, in 1886, he wrote: 'when we say the atomic weight of, for instance, calcium is 40, we really express the fact that, while the majority of calcium atoms have an actual atomic weight of 40, there are not a few which are represented by 39 or 41, a less number by 38 or 42, and so on.'

The actual situation turned out to be somewhat different from the one that Crookes proposed—chemical elements do not generally have two series of lighter and heavier isotopes with symmetrically reducing occurrences around their prime atomic weight. In addition, some elements have just one stable isotope, and others only two. And yet, taking into account that in 1886 virtually nothing was known about the atomic structure or the existence of nuclei, the above quotation gives a clear example of phenomenal intuition, comparable to the statement (published around 1815) by the physicist and chemist William Prout (1785–1850) that 'all elements are made of hydrogen', as indeed they may be said to be. This may be said to fall into the same category as the much earlier idea of Thales of Miletus (about 624–546 BCE) that 'all is made of water', which, as we know now, is a hydrogen oxide. When Ernest Rutherford (1871–1937) coined the word 'proton' to designate the prime building particle of any nucleus, he (as they say) may have been motivated by its similarity to the last name of William Prout. And (although it is rarely used) 'prout' is the unit of nuclear binding energy, which is equal to 185.5 keV (kiloelectronvolt), that is 1/12 of the binding energy of deuteron, the latter being 2.2 MeV.

However, it is not my intention to go into a detailed history of stable isotope science and technology as these aspects are well represented in the literature and online, I mostly (but not exclusively) confine the following discussion to what I believe are my own contributions in the area of isotopicity and isotopic randomness.

## 4.4 Isotopic curiosity and prime numbers

A little detour to some kind of metaphysics and number theory is due here. Both metaphysics and number theory are in my gallery of most beloved objects (none of them are 'objects' in a strict sense though…). As mentioned above, chemical elements have a varying number of naturally occurring stable isotopes. If $Z$ is the atomic number (that is, the place number of the element in the periodic table), the range of stable isotopes runs from hydrogen with $Z = 1$ to bismuth with $Z = 83$ (which is, actually, slightly radioactive; see section 4.1).

A glance at the table of isotopes shows that some elements have only one stable isotope, while their immediate neighbors (the next $Z$s) may have several. It all looks pretty random. The two most numerous groups are mono-isotopic and di-isotopic elements. They are almost equal in number (21 and 22 elements, respectively) and together encompass about half of all stable elements. As was mentioned in section 4.1, it is a curiosity of Nature that two elements inside the stable range of $Z$ do not have stable isotopes at all, namely, technetium ($Z = 43$) and promethium ($Z = 61$).

While the author is not aware of any universally accepted explanation for the existence of these two 'gaps', it is interesting to note that both of the above values of $Z$ (43, 61) are prime numbers. It may be just a curious coincidence, or there may be hidden physical reasons. However, taking into account the frequently emphasized role of prime numbers in various natural phenomena (Aczel 2000, Derbyshire 2004, Plichta 1997, Ribenboim 1989), one may refer (metaphorically, perhaps) to some peculiar numerological game that Nature is playing in this case. Whether this game will become more lucid through some deeper theories relating the physical universe and the structure of mathematics remains to be seen, but the discourse continues (e.g. Tegmark 2014).

A few more comments on prime numbers and isotopes are due here, in my view. The distribution of prime numbers and related areas of number theory (e.g. solutions to the Diophantine equations) form the basis of the unchangeable Platonic world in the sense that 'they are just out there'. We just discover (for ourselves) the numerological patterns but in no way 'invent' or 'create' them. This pattern of prime numbers is absolute, immutable and ever-the-same in any possible world and even in any simulated reality (Berezin 2006).

Nothing can change the distribution of (the infinite number of) prime numbers, not any 'higher reality' and not even 'God' (whatever images we can have for these ideas). Neither can 'He' change a single digit in the infinite decimal expansion of the pi number (3.14159...) or any other fundamental constant of mathematics (say, $e$, or the square root of two, etc). In fact, long philosophical traditions (Pythagoras, Plato and others) put some kind of an 'identity sign' between the notion of God(s) and the infinite unchangeable world of numbers and forms.

This is not quite so for the 'physical' laws of 'our' universe, which are somewhat provisional and contextual. Among various mathematical (Platonic) entities, prime numbers and their distribution are often seen as playing a special role (e.g. Plichta 1997). Likewise, human imagination, culture and folklore demonstrate the attribution of significance to prime numbers. This is quite remarkable and persistent, e.g. two is indicative of love/sex and duality, three is the trinity (in Christian theology and outside of it), seven is a traditional number for luck, five and 13 are used in the occult and 11 is at the center of the recent lore of 'elevenology' (11:11), which was initiated by a New Age 'priestess' with the (pen) name 'Solara' around 1990.

In fact, 11 is also the first 'twin prime' number; its 'twin' is 13. One might think that two, three, five and seven, which are all primes, also form two sets of twins, namely 3–5 and 5–7, however they are a kind of exception, the only case where three primes form a close triplet (or a quadruplet if we add two to their company), and 11–13 is the *first* 'isolated' twin prime. Curiously, the Canadian one-dollar coin (the 'loonie') has 11 corners (it is a regular 11-gon). As for higher primes, quite often 17 and 37 are met in various fables and stories, while the prime number 137 (the inverse of the thin structure constant in atomic physics) has almost a cult or mystical significance for some physicists.

It is interesting to note here that due to the universality of prime numbers they are (were) used for interstellar communications. The so-called Arecibo message that was radio-beamed to the cosmos in November 1974 as part of the Search for

Extraterrestrial Intelligence (SETI) experiments was based on prime numbers. The message was binary coded in 1679 pixels using the fact that 1679 is a semi-prime (the product of two primes, 73 and 23). If such a message is intercepted by some civilization (and any advanced civilization will, of course, know of prime numbers), they will factor it ($1679 = 73 \times 23$) and restore the original graphical picture that carries some information about our civilization. Of course, more complicated pictures and messages can be coded in a similar way using much bigger semi-primes (e.g. $36928907 = 4219 \times 8753$). Assuming that binary digits can be represented by two different isotopes (e.g. $^{12}C$ and $^{13}C$), such a coding in a two-dimensional (or three-dimensional) pattern can, in principle, be used for isotopic information storage systems (section 4.8).

What I venture to propose (as an original idea or not) is that the infinite intricacy of prime number distribution forms the basis for the universal connectedness of (arbitrary) distant points and events. This seems to resonate with the ideas of non-local quantum connectedness, such as quantum potentials and the 'holomovement' idea of David Bohm (Bohm 1952, Bohm and Hiley 1993), or recent (theoretical and experimental) developments regarding Bell's theorem (Albert 1992). To this we can add the ideas of the cosmic size of the wave function of the quasi-stationary states of radioactive isotopes (section 4.17).

Physical (and/or metaphysical) vacuum (in any imaginable model) is never 'empty', it inherently and unavoidably contains all the Platonic world and, correspondingly, an infinite capacity for informational unfolding and emergence. Different segments of prime number sequences may be 'responsible' (like different segments of DNA in genetics) for different aspects of the universal unfolding and concretization of particular sets of physical laws in individual baby universes of an inflationary cosmos and/or Everett's cosmogenetic chains.

What 'physics' really does in this picture is to 'label' some specific patterns from the numerological and primological 'Platonic world field' to specify a set of particular objects and/or modes of existence from the entire Platonic world. This 'mechanism' of 'specification' by 'materialization' (or 'embodiment') of the Platonic world is metaphorically similar to the reduction (collapse) of the wave function (Penrose (1994), (1996), Berezin (1992a), 1994c). 'Physically' informational connection (Platonic world decoding) through space and time can proceed by (any) mechanism of default labeling, e.g. by pattern proliferation mechanisms in nonlinear dynamics (informational cloning) or by (any) kind of pattern-cloning process, e.g. 'cosmic censorship', phase locking, morphic resonance, solitonic mode selection, etc.

As was mentioned above, two elements in the middle of the periodic table, technetium ($Z = 43$) and promethium ($Z = 61$), have no stable isotopes and both have prime atomic numbers (43 and 61 are primes). To continue on from this, I also note that the two stable isotopes of silver, $^{107}Ag$ and $^{109}Ag$, are both prime numbers (both 107 and 109 are primes), as is the atomic number of silver itself ($Z = 47$) (both stable isotopes of silver occur in nature in almost equal proportions). One may wonder if this 'love of prime numbers' (which the silver element apparently shows) has anything to do with the alleged mystical capacity of silver to fight dark forces, etc (folkloric silver bullets to slay werewolves, etc). And gold ($Z = 79$), which (unfortunately?) only has one stable isotope, $^{197}$Au, is also a 'prime number' element, as both 79 and 197 are prime numbers. Ancient civilizations were laden with religious and mythical beliefs, and silver and gold were believed to be favored by the gods, who kept the metals shiny and rust-free.

In voicing these alleged connections between isotopes, prime numbers and folkloric mysticism I should make another comment. It is prudent for me to mention here that it is only in my present status as a Professor Emeritus (retired professor), who does not need any more promotions and appointments, that I can put into print such 'New Age' comments, which almost certainly almost any mainstream scientist would gladly dump into a crazy box. But again, who can guarantee that there will be no takers for these ideas and possible connections, if not now, then perhaps 50 or 100 years hence? I am leaving this as an open challenge for readers and posterity.

## 4.5 Isotopic freedom and many facets of isotopicity

Here I want to introduce the concept of 'isotopic freedom' as arising from the interplay of isotopic randomness and isotopic correlations. Then I discuss the informational storage capacity hidden in the isotopic combinations in molecular and solid structures and after that review the various aspects of isotopic diversity relevant to the functioning of biological systems (chapter 5). In this regard it is worth observing some parallels between classical isotopic randomness and the 'usual' quantum mechanical indeterminism (Berezin 1987d).

Next, I put isotopicity into the context of the quasi-mental self-organizing processes found in Nature, especially in the realm of thermodynamically open non-equilibrium systems. We shall examine the extent to which isotopicity may be seen as a phenomenon that is compatible with thermodynamic openness and those effects commonly related to the mechanisms of spontaneous pattern formation through fluctuations, as discussed by Ilya Prigogine and many others (Nicolis and Prigogine 1977, Haken 1978). Later we look at some more specific analogies and connections between isotopicity and consciousness (Berezin 1990, 1992a, 1994a, 2015).

In chapter 5 we review the isotopicity–mind relationship, in terms of both the underlying physical mechanisms and some holistic-type mind–matter interactions (of the type that some claim occur in connection with water memory, homeopathy, crystal healing and the like).

## 4.6 My 'Newton's apple' of isotopicity

*Since I don't have any scientific reputation to lose, I can say what I please without giving a damn about what the professionals think of it.*
Arthur C Clarke (1917–2008)

The lives of many great people are surrounded by legends. One very well circulated one concerns how Isaac Newton discovered the law of gravity. As the story goes, young Isaac was sitting under an apple tree, suddenly an apple fell on his head, and in a minute he had come up with the idea of universal gravity. And right away, in a flash of an intuition, young Newton realized that the fall of an apple is caused by the same force that holds the Moon to its orbit. The rest, as they say, is history.

We can smile all we want at this story, but whether it was an apple or something else, the essence of this tale is almost certainly true. Most likely Newton—almost instantly—thought this way: if the force of gravity can reach the top of the tree, why not suggest that it can go much further, perhaps all the way to the Moon. And here he got the main idea: the force that makes an apple fall and the force that holds the Moon to its orbit is the same force. Click a mouse and the printout reads 'the law of universal gravity' (they say that Newton would easily have been awarded three or four Nobel Prizes for his work had they existed in his time).

So, what was my Newton's apple for my idea of isotopicity as a general conceptual framework for the effects of isotopic diversity? My story here is probably even a bit longer than it was for Newton. I can trace it to the work in solid state physics that I did while still in Russia (the USSR) in the early 1970s.

In 1970 I defended my PhD thesis (in Russian: my *Kandidat Fiziko-Matematicheskikh Nauk* dissertation) at the Department of Theoretical Physics, Leningrad University. My topic of study there was the quantum theory of the impurity centers in crystal lattices (color centers); I later published a few papers in this area (e.g. Berezin (1969), (1970), Berezin and Kirii (1969)) and continued with this topic, sometimes with experimentalists (e.g. Kashkai *et al* 1972a, Kashkai *et al* 1972b).

After defending my PhD dissertation, I found a research position at the Institute of Semiconductors of the Academy of Science, which later amalgamated with the world-famous Fiztech (the A F Ioffe Physical-Technical Institute of the Academy of Sciences) in Leningrad (now Saint Petersburg).

My work at that time at Fiztekh was a mixed blessing. On the one hand, there were all the typical adjustments of a first professional job, while on the other hand there was all the excitement, drive and enthusiasm of youth. Specifically, I worked at the Laboratory of Thermoelectricity, which was composed of some 50 or so people. I did my best to stay away from all the internal cat fights in the laboratory (almost any lab anywhere in the world has them) and worked mostly at home, coming to the institute as a rule once or twice a week (a kind of de facto privilege theoretical physicists usually had, although it was not a part of any official policy). The work I was assigned to concerned a theoretical study of the conductivity mechanisms of crystalline boron (called beta-boron). It was an amazingly interesting and challenging topic for me and it resulted in a few publications with experimentalists (e.g. Berezin *et al* 1972, Berezin *et al* 1973). However, when I moved to Canada and was working at McMaster University, this beta-boron turned out to be the bridge to my work on isotopes and the isotopicity concept.

Boron is the fifth element in the periodic table, its atomic number is $Z = 5$ and it has two stable isotopes $^{10}B$ and $^{11}B$, with an approximate ratio of one to five (Berezin 1984e). When I started working on boron, I did not pay attention to the fact that it has two isotopes, this was somehow out of my scope at that time. What is interesting and peculiar about beta-boron is that it has a very complicated crystal structure with an elementary cell consisting of 105 atoms with different coordination numbers. Normally, one would expect a complex crystal structure for compounds of several different chemical elements. However, crystalline boron is something of an exception in this regard—it has a very complex crystal structure based on fivefold (icosahedral) symmetry even in its elementary form (it is a single chemical element and not a compound with any other chemical element). In short, it was a fascinating material to study theoretically, I was truly excited about the work I was doing.

My more targeted focus was on the mechanism(s) of electrical conductivity in boron. Some ideas that were around at that time about the nature of electrical conductivity in boron were pointing to the presence in it of so-called hopping conductivity. The term hopping conductivity means that electrons move through a crystal lattice via consecutive hoppings (jumpings) between some localized (or quasi-localized) states in it. To form this type of conductivity the crystal needs some form of disorder to produce centers of electronic localization, such as the disorder introduced by randomly located impurity atoms or some other distortions. Such a disorder destroys the pure periodicity of the crystal lattice (the 'translational invariance') and that, according to quantum mechanics, can lead to the formation of trapping centers, and these are the regions where the electrons can temporarily reside (centers of quasi-localization). It is about the same as our ability to skip over puddles of water after heavy rain by jumping via some stones or dry spots. That is how you can envision the hopping conductivity process (Bottger and Bryksin 1985, Shklovskii and Efros 1984).

As such, hopping conductivity was found experimentally and studied extensively, experimentally and theoretically, for so-called doped semiconductors. The latter means semiconductors containing impurities. An example is Si(P) which is crystalline silicon with phosphorous atoms incorporated in a crystal lattice. In such systems electrons can hop (jump) between the impurities. But what about pure boron, which contains no significant concentration of impurities? What can cause the formation of centers of localization in this case?

In 1981, just a year after I joined McMaster University as an (associate) professor of engineering physics, I attended a conference on boron in Uppsala, Sweden, where I presented a paper (Berezin 1981a) on the electrical properties of boron. In the discussion of the possible cause of the hopping conductivity in boron somebody (I do not remember exactly who) made a passing remark that because boron has two stable isotopes ($^{10}$B and $^{11}$B) their random location may indeed account for the formation of centers of localization. This, in turn, could account for the hopping conductivity in crystalline boron.

Upon returning to Canada, when I was working on a paper for the follow-up journal issue for this conference, I mentioned this idea in just a few words (Berezin 1981b), where I discussed possible sources of disorder leading to electronic localization in boron. The quote from the said paper mentioned isotopic disorder among a few other options: 'in a crystal with a very complex unit cell, even a small perturbation (e.g. that is due to residual impurities, to dislocations, to thermal vibrations or even, as in boron, to a random distribution of the $^{10}$B and $^{11}$B isotopes among the lattice sites) is able to create the degree of disorder sufficient for the occurrence of the quasi-localization and hopping conductivity'.

Later I thought about isotopic randomness in this context more and that is how my general interest in isotopicity was awakened. Of course, I realize that the seed idea in this case was not entirely mine (who was this anonymous conference participant who made this passing remark?), and I believe that somewhat similar occurrences are quite common in science. Perhaps, sometimes 'big ideas' can indeed be triggered by some conversations over coffee or beer, when people may not even exactly know each other. Or maybe some other, totally random, inputs can put a researcher onto a line of thinking that may later take many years to mature into a cohesive major idea or a hypothesis.

As I mention elsewhere in this book, I do not much care about claims of priority, of who said this-or-that first. Maybe, somebody else thought of this possibility before me, maybe it was even published somewhere in the great quantities of now-almost-forgotten research reports and papers; yes, maybe. Maybe, apples also fell on the heads of people long before Newton and they also thought of universal gravity. Again, maybe. René Descartes said something like this: 'There is nothing so strange and so unbelievable that has not been said by one philosopher or another'. But whatever the case may be, personally for me, this passing remark about boron isotopes, $^{10}$B and $^{11}$B, was the launching pad for my isotopicity journey, which was bound to last for many years. That is how I came to have my own apple fall on my head.

## 4.7 Isotopic fiber optics

There are many possible ways in which the diversity of stable isotopes can mingle with physical effects at micro- and nanoscales. For example, isotopic randomness in solids and crystals affects their thermal conductivity (Klemens 1981). The relevant isotopic effects can be indicated for a variety of scientific fields, ranging from traditional solid-state physics to biology and geology. Selective use of specific isotopes to fabricate micro and nano-structures constitutes an area which can provisionally be defined as isotopic engineering (Berezin (1989a), (1993a), (1993b), Plekhanov (2004), Haller (1995), (2002), (2005), Cardona and Thewalt (2005), Itoh (2009), Itoh and Watanabe (2014), Itoh *et al* (1994), Watanabe *et al* (2009), Mukherjee *et al* (2015)). Some examples along this line are isotopic light confinement, isotopic fiber optics and isotopic information storage.

Let us imagine an isotopic interface, i.e. a boundary between regions with the same chemical identity but of different isotopic composition. Differences in the refractive index at both sides of isotopic interface could lead to the possibility of the total internal reflection of light and, consequently, could provide an alternative route for light confinement (Berezin 1988b, 1988c, 1989a). Isotopic fiber is a structure in which the core and cladding have the same chemical content but different isotopic composition. The boundary between isotopically different regions forms an isotopic interface (figure 4.2). This is a primary requirement for any fiber optics system. For a quantitative estimate, let us consider a boundary between $SiO_2$ (the main component of silica) where both sides are identical chemically and structurally but have a different

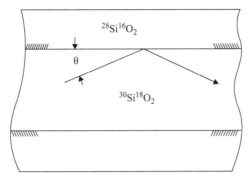

**Figure 4.2.** Isotopic fiber in which core and cladding are both pure $SiO_2$ (silicon dioxide) but with different isotopic composition, in the example shown the core is made of $^{30}Si$ and $^{18}O$ while cladding is from $^{28}Si$ and $^{16}O$ (other combinations stable isotopes are, of course, possible too—both oxygen and silicon have three stable isotopes each). Light rays hitting the isotopic boundary at angles less than critical ($\theta$) experience total reflection into the region of greater refractive index.

isotopic composition, for example, one side is made of $^{28}Si$ /$^{16}O$ and the other side is made of $^{30}Si$ /$^{18}O$. Isotopic separation technologies (e.g. Ishida 2002) can certainly provide the starting materials to fabricate such isotopic interfaces.

In isotopic fibers in which the core and cladding are made of different isotopes, the half-angle of the acceptance cone could be up to several degrees. The resulting lattice mismatch and strains at the isotopic boundaries are correspondingly one part per few thousand and, therefore, can be tolerated. Further advancements of this isotopic option could open the way for essentially monolithic optical chips with built-in isotopic channels inside the fully integrated and chemically uniform structure. The above idea of isotopic fiber optics was suggested in a few of my publications in 1988–89 (Berezin 1988b, 1988c, 1989a) and at the time of this writing (2016) I am unaware if any further theoretical and/or experimental uptake of this idea has been pursued.

## 4.8 Isotopic information storage (digital isotopicity)

As atoms, isotopes go one by one, and that fact alone makes them inherently digital entities. They can be 'read' as if they are letters in some atomic-scale alphabet. This fact prompted me to come up with the idea of isotopic information storage (Berezin 1984d, 1984h, 1986b, 1988e). While I do not know if anyone came up with the same idea before me, as may well be the case, all I can say is that I came up with this idea independently of any prior input. In the same way that neither Copernicus nor Herbert Spencer may have been the first to come up with the ideas of (respectively) the heliocentric system and biological evolution as survival of the fittest, nonetheless they both took the major credit for these ideas in the pages of history.

As was mentioned above, infinite periodical structures are much easier to describe mathematically than non-periodical and/or finite structures. From an informational point of view this means that non-periodical structures have a far greater capacity for carrying information than an ideal periodical (and hence a featureless) structure. A book whose pages are all filled with the letter A (AAAAA...) contains much less information (in fact, almost none) than a real book of meaningfully arranged characters.

Recall the image of a hypothetical library which contains all possible books—here I refer to the library of Babel, as discussed above (section 2.5). This (mystical and metaphysical) library (Borges 1998, Bloch 2008) contains books with every possible permutation of common printed characters and the size of such a library would be far, far greater than the size of the observable (Big Bang) universe (Bloch 2008). Can we re-project a similar image onto crystals with isotopic randomness? (Almost all crystals and other structures have plenty of isotopic randomness at the atomic scale.)

So, applying a similar argument for crystals with isotopic diversity, one cannot fail to observe that isotopically mixed crystals can potentially carry much more information than single-isotope lattices. Information can be coded in the positions of various isotopes within the crystal lattice. Or, perhaps, in some more holistic ('holographic') structures, such as, for example, those created by spatially extended magnetic fields which, in turn, are produced by the nuclear moments of isotopes (such as, say, $^{13}C$, which has non-zero nuclear magnetic moments).

Thus, isotopic differences can lead to novel systems of information storage at the nanoscale level. Isotopic information storage consists of assigning the information 'zero' or 'one' to monoisotopic micro-islands (or even to single atoms) within a bulk crystalline (or thin film) structure. This technique, if developed, could lead to a very high density of information storage of up to $10^{20}$–$10^{23}$ bits cm$^{-3}$. This is because isotopic information storage (unlike other information storage techniques) allows the information bit to be carried by a single atom.

One can estimate that the information content of the Library of Congress is $10^{17}$ bits (estimate: $10^8$ volumes of 1000 pages, each with a generous allowance of $10^6$ bits per page—the latter to account for the digitized photographs). This means that the entire content of the Library of Congress can be isotopically stored in one cubic millimeter. Of course, proper three-dimensional methods of writing and reading are required for that. These can be developed along the lines of atomic force nano-technology, which allows the manipulation of individual atomic species.

Furthermore, the main potential advantage of isotopic information storage lies in the fact that the information is incorporated in a chemically homogeneous matrix. There are no chemically different impurities (like those existing in optical storage with color centers) or grain boundaries between islands of drastically different magnetization (which is a limiting factor in common magnetic storage techniques). Information stored in an isotopic recording exists as a part of a regular (in principle, ideal) crystal lattice. By 'ideal' we mean here that the structure does not need to contain any of the common defects inevitable in heteroatomic coding and as such isotopically stored information is protected by the rigidity of the crystal lattice itself.

It should be noted here that an off-set effect may exist, which could be a potential problem for isotopic information storage that uses the positions of individual isotopes. Namely, there is the (potential) possibility of neutron tunneling (isotopic castling), as is discussed in chapter 5. This random isotopic hopping (or, alternatively, neutron tunneling between nuclei) can damage (or even destroy) the information (digitally) stored in the positions of isotopes.

Thus, for isotopic information storage systems, methods to suppress this effect of isotopic hopping should be developed in order to protect the stability of the information storage. For example, instead of using a single isotope as a digital '0' or '1', it is possible to use complexes (clusters) of, say, ten or 100 isotopes to store a single bit. While this, theoretically, correspondingly reduces the density of the information storage by the same factor (by how many isotopes form a cluster), it will still allow for very high absolute densities of three-dimensional information storage. In practical terms, it does not seem to be such a big difference if the entire Library of Congress can be stored in one cubic centimeter, rather than in one cubic millimeter.

In crystals, the distance between atoms is about three angstroms (1 cm $= 10^8$ angstroms). This gives about $10^{24}$ atoms per cubic centimeter. Assuming that one binary (0 or 1) can be supported by an isotopic structure cluster of, say, 100 atoms inside the crystal lattice, we can estimate that for three-dimensional holographic-type information storage one cubic centimeter of isotopic memory chip can store $10^{21}$ bits of information. With the above estimate for the bits in the Library of Congress ($10^{17}$), this comes to about 10 000 such libraries in one cubic centimeter. While I do not know if (and when and where) such a chip can be created, I venture to suggest that it be named after Jorge Luis Borges, the author of the *Library of Babel* (see section 2.5). Borgession, perhaps.

## 4.9 Isotopic superlattices

It looks to be almost a trivial statement to say that elements with two or more stable isotopes should form a random distribution over the regular sites of a crystal lattice. This is usually taken for granted because isotopic differences in lattice binding energies are considered to be too negligibly small to affect the dynamics of lattice formation at, say, crystallization from melt.

Isotopic disorder in solids affects the whole variety of thermal, optical and electrical phenomena. The most profound effect of isotopic randomness is probably the modification of lattice thermal conductivity. Indeed, some studies show that in some cases there is a significant difference in thermal conductivity between the natural (isotopically mixed) and isotopically purified samples (Klemens 1981). This effect is attributed to the isotopic phonon scattering that can be treated as a separate scattering mechanism (Berezin 1984f, 1992a). Similar (although somewhat weaker) isotopic scattering could affect the mobility of charge carriers (Berezin 1984f) and in some extreme cases could lead to electronic localization in narrow conduction bands (Berezin 1984e).

Consider a defect-free periodical crystal lattice which, however, has some large-scale structure of isotope distributions (figures 4.3 and 4.4). Generally speaking, one could think of some mechanisms leading to the spontaneous isotopic ordering and even to the formation of isotopic superlattices (Berezin (1987a), (1988a), (1988d), (1988e), (1989b), (1990), (1992a), Kojima *et al* (2003), Bastian *et al* (2010)).

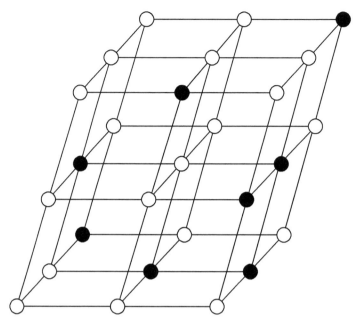

**Figure 4.3.** Crystal with two (or more) stable isotopes may be perfectly symmetrical and ordered in terms of the geometrical structure of its crystal lattice, yet, generally, it has randomness in the arrangement of various isotopes over the lattice sites (examples: $^{12}C$ and $^{13}C$ in diamond, or $^{10}B$ and $^{11}B$ in boron crystal; cubic lattice shown as a simplest example—most crystals have more complicated symmetries of a lattice). In this figure 4.1 isotopes (white and black dots) are located randomly at the regular lattice sites.

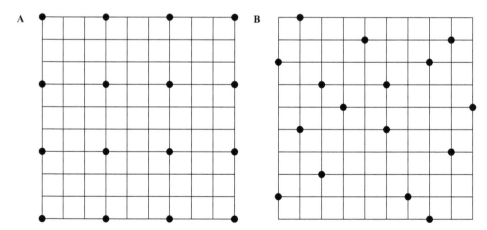

**Figure 4.4.** Two dimensional illustration of isotopic randomness and isotopic superlattice. In (A the) 10 × 10 lattice has two isotopes located randomly, while in (B) the minority isotope forms a superlattice. These are two extremes: in actual cases there may be a continuous range of partial isotopic ordering with partially developed isotopic superlattices.

This could be similar to the known ordering of voids under irradiation. It is easier, however, to think of such a non-uniformity as being created artificially within the otherwise perfect (i.e. free of ordinary defects) crystal lattice. The creation of such isotopically ordered structures is obviously a technically achievable task at this stage (e.g. Haller 1995, 2002).

Some heterogeneous systems, as a rule, exhibit a trend towards a variety of segregation and ordering scenarios for a variety of physical situations (Berezin (1987a), (1988a), (1988d), (1988e), (1989b), (1990), (1992a), Goldman and Berezin (1995), Ibrahim and Berezin (1992)). The isotopic shifts of vibrational frequencies in anharmonic crystal lattices result in isotopic variations of the lattice constants which (under specific crystallization conditions) may lead to positional isotopic correlations and, in extreme cases, to the formation of isotopic superlattices.

## 4.10 Isotopic quantum wells

Isotopic non-uniformities can originate from a variety of phase-separation non-linear processes. For concreteness, let us consider an isotopically pure crystal (e.g. $^{28}$Si) inside of which there is a spherical inclusion of $^{30}$Si . We use a sphere here just to simplify the estimate; the particular shape of this 'isotopic island' is, in fact, unessential.

Estimates show that the above isotopic inclusion could work as a potential well for electrons (Berezin 1987b, 1987d). It is known that the lattice constants of isotopically pure crystals are slightly different. The observed differences are usually of the order of one part per thousand. These isotopic variations of lattice constants are explained by the differences in bond lengths for various isotopic pairs (e.g. for $^{28}$Si–$^{28}$Si and $^{30}$Si –$^{30}$Si in Si crystals). The latter, in turn, are related to the anharmonicity of zero-point vibrations.

The above-described isotopic inclusion produces a lattice constant mismatch of the order of one thousandth (1/1000). This, in turn, will result in some strains that are qualitatively similar to those of the currently discussed case of strained super-lattices. Although the magnitude of the effect is much smaller than in the case of chemically different atoms (e.g. Si/Ge superlattices), it is, nevertheless, non-zero. Assume that in a certain isotopically mixed crystal (e.g. in Si) one isotope is replaced by another (e.g. $^{28}$Si by $^{30}$Si) in just one lattice site. In view of the above, such a change will result in adjustments of the equilibrium positions of the nearest neighbors. These shifts can be of the order of $d/1000$ ($d$: inter-atomic spacing) and are equivalent to the presence of random strains that vary from site to site. The variation of bond lengths in isotopically mixed lattice can now be equated to the presence of some randomly varying (because of the random location of isotopes) central forces emanating from each atom.

Consider now the net (i.e. subtracted) interaction between the two atoms of the same isotope, e.g. between two $^{28}$Si atoms in a $^{30}$Si host lattice. Depending on a particular case, this net effect can have either sign, i.e. it can be either attractive or repulsive. The first should generally favor isotopic precipitation, and the second isotopic ordering (same isotopes tend to avoid being close neighbors). We should not, therefore, expect the ordering tendency in all cases. However, the character of this resulting net interaction between minority isotopes in a majority matrix can be regulated by a particular ratio of the isotopes (i.e. a 99% to 1% mixture of $D_2O$ and $H_2O$ will apparently behave oppositely to a 1% to 99% mixture of the same).

To estimate the size of the isotopic fluctuation needed for the formation of the bound state, we can use the Kronig–Penny three-dimensional model of zero-strength delta potentials (Berezin and Kirii (1969), Berezin (1973), (1986a), Demkov and Ostrovsky (1975)). This model contains only one parameter: the lattice spacing $d$. Each lattice site (atom) is replaced by a three-dimensional delta-potential well with a zero value strength constant (regarding the latter, in an asymptotic sense the strength constant tends to +0). In physical terms this means that the bound state for such an isolated well has an arbitrary small binding energy (+0). It is known that two (or more) delta wells of zero strength separated by a finite distance have a bound state with a finite (that is non-zero) binding energy, even if each such delta potential, taken individually, is unable to support the bound state (Berezin 1982b, 1984e, 1986a). This may appear paradoxical, yet this is a purely quantum mechanical effect.

## 4.11 Isotopic tribology

Tribology is an area of science and engineering that is related to the effects of friction between two surfaces sliding upon each other. While at the macro level friction is an important area of engineering and material science, some recent advancements have also been made with regard to surfaces at micro- and nanoscales. In the common technological context, the control and reduction of friction effects is usually done by some form of lubrication (oil, etc). Although in the engineering context the friction effects at contacts are usually discussed in the realm of purely classical physics and mechanics, at the microscopic level quantum behavior can become significant and in some situations it can become dominant. One particular example of this is the quantum lubrication effect (Feiler *et al* 2008, Munday *et al* 2009, Cranston and Gray 2006, Lamoreaux 2009) and possible modifications to it arising from isotopic randomness and isotopic structuring (Berezin 2011).

Among the interesting quantum effects taking place at the nanoscale is the so-called repulsive Van der Waals effect. Van der Waals forces, also known as dispersive forces, can be attractive and repulsive. The repulsive Van der Waals effect, also known as the repulsive Casimir–Lifschitz effect (Feiler *et al* 2008; Munday *et al* 2009; Cranston and Gray, 2006; Lamoreaux, 2009), is not exclusively related to isotopic diversity. It is based on quantum electrodynamics and its explanation involves pressure effects from the zero-point vibrations of virtual electromagnetic fields in the contact region. The effect goes on the differences of the dielectric constants of both surfaces and the thin layer of liquid medium between them.

Several other isotope-related effects can also likely play some role in the formation of the above-discussed repulsive forces. For example, nuclear magnetism, being an isotopically selective effect, can play an especially prominent role at the contacts of two surfaces (quasi-two-dimensional systems) through the formation of a network of magnetic moment interactions. The latter resembles artificial (spin-glass) neural networks (Berezin 1992a) and can enhance energy exchange processes. This, in turn, may affect (reduce) the friction between sliding surfaces acting in analogy with the repulsive Van der Waals forces, leading to a quasi-lubricating effect (where the 'lubrication' is provided by quantum interactions rather than some specific material ingredient).

In this regard, isotopic engineering can possibly amplify quantum repulsive mechanisms. No experimental verification of this alleged isotopic effect is known to this author at this point. It would be of special interest to relate these effects to the earlier studies on quantum bound states in repulsive potentials by Neumann and Wigner (von Neumann and Wigner 1929). They gave an explicit example of a repulsive potential that nonetheless has a bound state. Such a paradoxical occurrence appears to be at odds with standard quantum mechanics, yet von Neumann and Wigner gave a neat interpretation of it as a 'reflection of a particle from infinity'. In this regard, isotopic structuring, as a tool for fine quantum adjustments (due to the general weakness of isotope effects), can serve as a way to form such refined situations when the quantum states contributing to quantum lubrication are becoming metastable (quasi-stationary).

Isotopic symmetry is an additional factor that can be introduced through proper isotopic engineering (e.g. the symmetrical patterns of isotopes at the surface). Such specific quantum effects as 'accidental degeneration' (*Zufallige Entartung*), may turn out to be both theoretically interesting and practically useful. In quantum physics degeneration traditionally signifies the situation where several distinguished quantum states have the same energy. Additional symmetry (in this case due to symmetrical isotopic patterning) increases the level of degeneration and makes it 'accidental' (Berezin 1992b). In the opinion of this author, the quantum effects in surface tribology (including quantum lubrication) are largely unexplored and potentially constitute a promising area of research and development.

Another class of largely unexplored possibilities may be related to energy transfer and electron hopping effects between nanoscale isotopic clusters at the interface region (Berezin 1987d, Goldman and Berezin 1995). Resonance energy transfer between clusters (or isotopic fluctuations) can affect the inter-surface tension and change (reduce or increase) the effects of quantum lubrication. Possible areas of application are, again, most likely related to small-scale devices like MEMS and bio-medical electronics. While the application of quantum physics to surface and contact mechanical effects is in its incipient stages, some advanced models of quantum lubrication have recently been discussed (Feldmann and Kosloff 2006).

It should be pointed out that our examples of isotopic effects in contact physics and quantum lubrication phenomena bear a largely heuristic character, indicating possible new directions of research. It is fair to say that not all of these examples may be actively pursued to further development and, conversely, some new, as of yet unexpected, isotope-related aspects may come forward.

## 4.12 Isotopic effects in corrosion

Physical phenomena occur in space and time. Although this trivial observation is generally true for any physical process, the description and modeling of various phenomena, like the disintegration of material and structures (Eberhart 1999, 2003) differ in emphasizing their temporal and/or spatial aspects. Some phenomena (e.g. phase transitions) can mostly be understood in terms of corresponding elementary steps, i.e. for them the 'historical' (evolutionary) aspect is relatively unimportant. For some other processes a more-or-less profound understanding requires their visualization in terms of their entire development, rather than as a mere sequence of elementary steps. The latter processes encompass such different phenomena as crystal growth, geological patterning (e.g. sedimentation), the growth and aging of single organisms and bio-evolution as a whole. For such processes holistic and descriptive models like catastrophe theory can provide integrative imagery (De Sa and Berezin 1989, Berezin 1991).

Corrosion is a multi-faceted physico-chemical process with elements of non-linear behavior (Berezin 1993a, 1993b). Phenomenologically, it exhibits a number of distinguishable scenarios, such as pitting corrosion, crevice corrosion, stress corrosion cracking, corrosion fatigue, inter-granular corrosion, biological corrosion, etc. Despite this diversity, the majority of the underlying microscopic mechanisms have certain common features. Namely, one can almost always indicate some non-linear dependences of the characteristic rate parameters (e.g. dissolution currents) on the concentrations of reagents, applied external electrical biases, etc. As such, the corrosion process is usually seen as an interplay of several chemical processes (anodic, cathodic, acid–base), which, as a rule, have various feedback loops that result in the formation of the overall non-linear diffusion–reaction system. Thermodynamically, an interactive system of 'metal plus active surrounding' is a non-equilibrium system. Therefore, the resulting situation can be commonly classified as a spatially extended non-equilibrium non-linear system. Such systems are generally prone to all kinds of fluctuations and isotopic fluctuations may play some role in pattern formation dynamics.

An interesting and scientifically challenging point here is the possible role of isotopic fluctuations in corrosion dynamics. According to the line of thinking advocated in this book, isotopic diversity (isotopicity) may have some (perhaps subtle and tangential) relevance to the physical aspects of corrosion (Berezin 1993a, 1993b). The role of isotopic diversity in corrosion initiation and propagation can be traced along two lines: (1) isotopic fluctuations as corrosion seeds and (2) the informational and pattern-forming aspects of corrosion. In terms of corrosion patterning we can even talk (in a somewhat metaphorical way) of the 'creativity' of isotopic fluctuations as pattern-forming factors. Experiments on corrosion rates on isotopically purified samples (as well as on samples with shifted isotopic abundances) could be one route to verify the relative scale of such effects.

## 4.13 Isotopes in quantum computing

The rapidly unfolding area of quantum informatics and quantum computing is an impressive growth industry at the frontier of current physics and electronics. While the basic principles of quantum computing are becoming generally known, the technological applications are still mostly in their infancy. Some authors go as far as ascribing (perhaps somewhat metaphorically) the enormous potential of quantum computers to their ability to borrow computing power from parallel universes (Deutsch (1997), Lloyd (2002), (2006)). In spite of the obvious speculative flavor of these ideas, many people find them fascinating and mind boggling.

At the time of writing (2016), there are several research lines attempting to implement quantum computing in practice. Some of them use the nuclear spin states of specific isotopes in crystalline matrices. In practical implementations on the basis of solid-state structures, quantum computing is isotopically selective. Thus, quantum computing naturally falls into the domain of isotopic engineering (Itoh and Watanabe 2014). Specifically, because of the quantum identity (indistinguishability) of the same isotopes (as opposed to the quantum distinguishability of *different* isotopes of the *same* element), isotopicity provides a natural playground for the establishment of quantum entanglement among large clusters of atoms. The latter (sustained quantum entanglement) is one of the key requirements needed for a functional quantum computer.

One of the most acute problems facing quantum computing is the need to maintain the quantum-coherent superposition of a system for sufficiently long (often macroscopic) times. To attain that, the quantum system should be sufficiently isolated from thermal baths and other stochastic perturbations. The latter perturbations produce decoherence and, hence, interrupt the process of quantum computation. For that matter, the nuclear spins of some specific isotopes may turn out to be systems of choice (nuclear spins are reasonably well insulated from the said perturbations). A number of concrete realizations can be tested for that matter. For example, small atomic complexes of isotopes with non-zero nuclear magnetic moments encapsulated inside fullerene balls may be reasonably well protected against outside interactions leading to decoherence.

## 4.14 Isotopic random number generators

*Anyone who attempts to generate random numbers by deterministic means is, of course, living in a state of sin.*

John von Neumann (1903–57)

Assuming a perfect (or almost perfect) isotopic randomness in solid and quasi-solid structures, individual counting of isotopes atom-by-atom can be utilized for the creation of nanoscale random number generators. Typically, the generation of random numbers in computers is based on the mathematical procedures of the truncation of various functions. Random numbers which are produced this way are, in fact, pseudo-random: due to the deterministic character of the computer codes, the strings of random numbers are repeated every time the same seed numbers are used. A mixture of stable isotopes, provided that it can be probed at the atomic level, is free from this limitation. An alternative way to use isotopes for random number generation is to use the isotopic jets produced by gas streams (Berezin 1987c).

Under the assumption of the existence of functional nanoscale reading technology capable of determining the isotopic identity of individual atoms (e.g. isotopically adjusted atomic force microscopy, AFM), it may be possible to produce strings of physically random numbers by scanning crystal surfaces. By counting different isotopes of the same chemical element (say, carbon) as digital 0s and 1s, a genuinely random non-repeatable binary string could be generated. This would include a proper normalization (scale adjustment) of the produced strings to account for the relative isotopic abundances. Like the random strings produced by time-clipping of the individual decays of radioactive isotopes, the strings produced by counting stable isotopes are free from the hidden correlations which are typical for the strings produced by algorithmic methods on computers.

## 4.15 Isotopic randomness as symmetry breaking factor

*Like the ski resort full of girls hunting for husbands and husbands hunting for girls, the situation is not as symmetrical as it might seem.*
Alan Lindsay Mackay (b 1926), British crystallographer,
Birkbeck College, University of London

The simplest conceptualization of the crystal lattice is a model of an ideal infinite periodical structure of ever-repeating identical unit cells. Such a theoretical construct is not only infinite, but by definition it has no defects or departures from ideal periodicity (the ever-sameness of unit cells). This image is at the core of standard theoretical methods and concepts, such as the Kronig–Penney model, $k$-vector diagrams, Bloch functions, Brillouin zones, electronic band calculations, etc. All these are consequences of the translational invariance (translational symmetry) of an ideal periodical structure (Berezin 1992a, 1992b).

Of course, in reality all crystals are of a finite size (actually, often very small) and have defects in their lattices (vacancies, impurities, dislocations, etc). And yet, theoretically and conceptually, it is easier to work with infinite ideal systems than with real and finite ones. Similar 'paradoxes of infinity' are known in mathematics, for example many functions (e.g. Gaussian exponent) are easier to integrate from zero to infinity than to a finite limit. Mathematically, it is far easier to describe all integer numbers (1,2,3,4,5,6,7, ...) than a smaller subset of only prime numbers (2,3,5,7,11,13,17,19,23, ...). The program (algorithm) for the set of all numbers is trivial, while the program that types out only prime numbers is much more complicated. And in the case of crystals, an infinite lattice model avoids the necessity of dealing with boundary conditions at the surfaces—the latter makes the mathematical analysis of finite structures with boundaries much more complex than that for infinite periodic structures. Speaking metaphorically, 'infinity is simpler than (any) finiteness'.

Now, imagine a defect-free infinite crystal lattice, for example, a diamond (pure carbon) or a silicon dioxide ($SiO_2$). Carbon, silicon and oxygen all have more than one stable isotope (C has two: $^{12}C$ and $^{13}C$, while Si and O have three each: $^{28}Si$, $^{29}Si$ and $^{30}Si$ and $^{16}O$, $^{17}O$ and $^{18}O$). Even under the assumption of the complete absence of any 'regular' lattice defects, these crystal lattices are not, strictly speaking, periodical. Different isotopes of each element will be located randomly and, hence, each particular atom will have a different isotopic environment at the atomic (nano) scale. In this regard, a random location of isotopes is a ground assumption (null hypothesis). An alternative option—the possibility of isotopic ordering (isotopic superlattices) is a separate issue, as discussed above (section 4.9).

Therefore, like regular crystal defects (impurities, vacancies, etc), isotopic randomness results in a violation of (or at least rather weak) translational symmetry. Generally speaking, a violation of symmetry (broken symmetry) has numerous consequences in physics, from elementary particles and cosmology to various aspects of condensed matter physics.

Thus, the question is, to what observable effects can the random location of isotopes in an (otherwise ideal) crystal lattice lead? Such alleged effects are most likely akin to the effects that result from the randomness of regular crystal defects. One such possible effect is Anderson localization due to isotopic disorder, which was discussed by this author earlier (Berezin 1982b, 1984e).

## 4.16  Instability of isotopically mixed systems

*We cannot control atomic energy to the extent which would be of any value commercially, and I believe we are not likely ever to be able to do so.*
Ernest Rutherford (1871–1937), speech to the
British Association for the Advancement of Science (1933)

*The phenomenon of radio-activity lead us straight to the problem of releasing the inner energy of the atom [...] The greatest task of contemporary physics is to extract from the atom its latent energy—to tear open a plug so that energy should well up all its might. Then it will become possible to replace coal and petrol by atomic energy which will become our basic fuel and motive power.*
Leon Trotsky (Lev Davidovich Bronstein, 1879–1940),
speech 1 March 1926

The two quotations above, which were spoken at about the same time and came from two individuals who are world-renowned (for quite different reasons), could not be more diametrically opposed to one another. The great physicist Ernest Rutherford, the discoverer of the atomic nucleus, expressed his disbelief in nuclear energy just a few years prior to the first nuclear reactor. And the second quotation, if the signature were removed from it, could well be thought of as coming from someone like Enrico Fermi, who was among the first designers of the first working reactor. And a fairly similar collection of contradictory quotations showing all grades of pessimism and optimism can be found nowadays on the issue of so-called 'cold fusion'. In this book, I am not going to go into the cold fusion controversy at any length, but I would like to provide some thoughts on this issue from the general view of thermodynamics and the quantum physics of quasi-stationary states.

Speaking thermodynamically, an isotopically mixed system is almost always a non-equilibrium system that is in some kind of an excited (metastable, or quasi-stationary) state above its true ground level. Therefore, such a situation can be commonly classified as a spatially extended non-equilibrium non-linear system. In fact, almost anything we see around us (including ourselves) is an/are 'isotopically mixed system(s)' capable, in principle, of the release of energy through nuclear rearrangements. For example, it is energetically possible that the reaction of an isotopic rearrangement such as

$$^{17}O + {}^{17}O \rightarrow {}^{16}O + {}^{18}O$$

would go spontaneously from an $^{17}O$ nucleus to another $^{17}O$ nucleus through neutron tunneling (Jiang and Berezin 1998). This, again, is an example of spontaneous symmetry breaking when a symmetrical system (both atoms are 17) transforms itself spontaneously to an asymmetric one (16 + 18). On the basis of the total binding energy of these nuclei it is easy to show that such a reaction is exothermic (with a release of extra energy). The excess of energy released in such reactions can, in principle, dissipate to low-grade forms (such as crystal lattice vibrations) instead of being emitted as a high-energy gamma photon.

Tables of stable isotopes are among standard physics data and are available in many forms. They normally list the stable isotopes of all elements, along with the total energies of each isotopic species. For example, for the three stable isotopes of oxygen, $^{16}O$, $^{17}O$ and $^{18}O$, the total energies (in the atomic unit of mass, amu) are 15.994915, 16.999132 and 17.999160, respectively. (For atoms the atomic unit of mass is defined as 1/12 of the mass of the $^{12}C$ atom, for which the energy equivalent is 931.5 MeV, with MeV being the megaelectronvolt.) It is easy to calculate the energy release in the above reaction as 0.004189 amu or 3.9 MeV, quite a large energy at the atomic scale.

To express this value in a more accessible way, it is easy to estimate that one liter of $^{17}O$ water (all the oxygen atoms in it are $^{17}O$) can (at least, in principle) produce, because of the above reaction, an energy release of $2 \times 10^{13}$ J, which is equivalent to burning about 650 m$^3$ of gasoline. The corresponding energy release of just 1 mL (milliliter = 1 cubic centimeter) of $^{17}O$ water is $2 \times 10^{10}$ J —an energy equivalent to the kinetic energy of 20 jet aircraft (the kinetic energy of one jet is about $10^9$, one billion joules). Likewise, the reader can easily calculate what the energy equivalent of a standard bar shot (30 mL, an ounce) of $^{17}O$ water would be.

To make this estimate even more impressive we can look at all the water on Earth. The total volume of the Earth's hydrosphere is about $1.3 \times 10^9$ km$^3$. That is 1.3 billion cubic kilometers, or $1.3 \times 10^{21}$ liters. Thus, taking into account the fact that the natural abundance of the $^{17}O$ isotope is 0.039% (one atom of $^{17}O$ per 2560 of all oxygen atoms), we can estimate that if all the water on Earth went through the reaction given by the above equation, this would amount to a total energy release of some $10^{31}$ J. For comparison, the total annual global energy used by humans is about $5 \times 10^{20}$ J. And the above estimate ($10^{31}$ J) refers only to the above reaction with oxygen isotopes. The total energy of the hydrosphere that could be released by hydrogen and deuterium fusion is estimated to be $10^{34}$ J. This would constitute about a million years of our present energy needs, although over that time (if we survive) we will probably manage to use the huge hydrogen balls in our solar system (Jupiter, Saturn, etc) and, perhaps, beyond it.

Such a reaction can occur spontaneously because of the energy minimization principle, no matter how 'difficult' it may be for a neutron to tunnel from one $^{17}$O nucleus to another $^{17}$O nucleus. However, in physics there are cases when the energy needed for a particle to tunnel through a potential barrier can be obtained from other sources in the system, for example, a multi-center Auger effect (Berezin 1969, 1983a) or stray cosmic particles that can bring in extra energy. And once such extra energy comes to the system, the other atomic pairs (say, $^{17}$O + $^{17}$O) can go through the same reaction via a domino effect.

Or, perhaps, some kind of Maxwell's demon (Leff and Rex 1990, Ehrenberg 1967) could be a convenient delivery service for it, as it can play well with the law of entropy. Or, alternatively, perhaps, some clever manipulation of quantum non-localities could enhance the tunneling probability from (tower) exponentially small to (almost) macroscopic ranges. The latter reactions may happen at a human timescale, as was suggested by Luis Kervran in his work on alleged biological nuclear transmutations (Kervran 1972).

Similar acts of neutron tunneling could be responsible for the nuclear trans-mutation effects in palladium-based systems (like Pd–D), which were used in what was claimed to be cold nuclear fusion (Fleischmann and Pons 1989 and later claims). Some research activity in this controversial area of low-energy nuclear reactions is still taking place in a few isolated places. The author of this book takes no personal position on this issue, considering that, in his opinion, the theoretical arguments in favor of and against this easy-to-obtain cold fusion at the present level of develop-ment are approximately of the same weight.

The above-described reaction of spontaneous neutron tunneling in the system of oxygen isotopes is just one of many possible reactions of this kind. If a chemical element has several stable isotopes (as many do), then there is always a lowest energy combination that can (at least in principle) be reached through neutron hopping from one nucleus to another. Such a possibility exists for any element with at least three stable isotopes with consecutive atomic numbers (e.g. O, Mg, Si, etc).

For example, for magnesium (stable isotopes are $^{24}$Mg, $^{25}$Mg and $^{26}$Mg), isotopic pairs like $^{24}$Mg–$^{26}$Mg and $^{25}$Mg–$^{25}$Mg are mutually convertible into each other through a single neutron tunneling jump. Depending on which of the two pairs has the lower total energy, the reaction $^{24}$Mg–$^{26}$Mg $\leftrightarrow$ $^{25}$Mg–$^{25}$Mg is exothermic in either the right or left direction. In either of these cases the reaction can be effected by a single neutron tunneling—either from $^{26}$Mg to $^{24}$Mg, or from one $^{25}$Mg to another $^{25}$Mg. For elements with an even greater number of stable isotopes many more combinations are possible and some may involve more than two participating centers. This is physically similar to the correlated electron tunneling in poly-center systems of impurities in crystals, the theory of which was developed by this author in the mid-1980s (Berezin (1984a), (1984b), (1984c), (1984d), Berezin and Jamroz (1984)). In spite of the fact that direct quantum-mechanical calculations for the tunneling of neutrons between nuclei give a negligibly low probability, an exponential (or even super-exponential) enhancement of quantum probabilities cannot be excluded, since no violation of the energy conservation law happens in such cases.

## 4.17 The cosmic scales of nuclear wave functions

*If the Sun is 1 mm in diameter, the closest star (4.3 light years away) will be 24 km (15 miles) from the Sun.*

Trivia (anybody can calculate this)

An interesting and perhaps challenging comment may be appropriate in relation to almost-stable (weakly radioactive) isotopes, such as bismuth-209 ($^{209}$Bi) or germanium-76 ($^{76}$Ge). These isotopes are usually considered to be stable because their lifetimes are of an order of $10^{19}$ to $10^{21}$ years, which is billions of times longer than the presumed age of the (Big Bang) universe (some 14 billion years). And yet, technically, these nuclei are radioactive and hence their ground state is, strictly speaking, a quasi-stationary state. The latter states belong to a continuous energy spectrum in terms of standard quantum mechanics.

Furthermore, because all matter is eventually radioactive, quantum mechanically all stable states are quasi-stationary states (QSSs). For QSSs, the asymptotic behavior (fall off) of (the spatial part of) the wave function at large distances $r$ can be approximated by the equation (Baz *et al* (1966), p 169, Landau and Lifshitz (1963), p 592):

$$\psi(r) = (1/r) \cdot \exp\left[(\tau/TL) \cdot r\right] = (1/r) \cdot \exp(r/R),$$

where $\tau$ is the characteristic nuclear time, $T$ is the lifetime of an isotope and $L$ is the characteristic length of a quantum confinement. (In this case $L$ is the 'size' of the nucleus in a quantum mechanical sense, which may not be exactly the same as its formal geometrical size, but is (usually) of the same order of magnitude).

The combination of the factors $TL/\tau$ is designated $R$. According to the formal reading of the above equation, at the characteristic distances of an order of $R$ the (spatial part) of the wave function starts to grow exponentially and for $r$ tending to infinity the amplitude of the wave function tends to infinity too. As an estimate (given a few lines below) shows, the value of $R$ is typically astronomical lengths (light years). The integral of the square modulus of the wave function given by the above equation is, of course, divergent, as is always the case for quantum states which belong to the continuum spectrum (and QSSs in a strict quantum mechanical sense belong to a continuum spectrum). The seemingly paradoxical result here is that the wave function in the above equation falls off so slowly that most of the wave function exists in the area outside the sphere with a radius of many light years.

This whole situation with unlimited growth of the wave function when $r$ tends to infinity may appear paradoxical. However, as Landau and Lifshitz noted

(Landau and Lifshitz 1963, p 592), the QSS is actually a non-stationary (decaying) state. Hence, the full wave function is exponentially decaying like $\exp(-t/T)$, which, however, does not negate the fact that the normalization integral of the wave function diverges (because $\exp(-t/T)$ remains finite at all $t$, here $T$ is the lifetime for the decay of the particular isotope).

The paradoxical result represented by the above equation can be interpreted as indicating that decaying QSSs (in this case, radioactive isotopes) do actually 'exist' (for the most part, at least) well 'outside' their formal locations 'here', and indeed some astronomical distance from 'here'.

As a specific example of the above, let us take the $^{14}$C isotope. It has a lifetime ($T$) of about 5000 years against beta-decay with a decay energy of $E = 0.156\,\mathrm{MeV} = 2.5 \times (1/10^{14})$ J. Taking $L = 1/10^{14}$ m (the size of the nucleus) for its size, and $0.42 \times (1/10^{20})$ s for the characteristic nuclear time, we obtain $R = 3.75 \cdot 10^{17}$ m, which is about 40 light years. Yes, strange as this result may look, 'most' of the $^{14}$C atoms we have around us are (according to this result) somewhere beyond Sirius (eight light years from us) or Vega (26 light years away).

For slower decaying isotopes (and according to the above, *all* radioactive isotopes are in QSSs) with greater values of $T$ (and hence $R$) this logic leads to even more remote regions for the 'prime residence' of these isotopes (or, rather, their QSSs). These may be thousands, or millions, or billions of light years from us. For such isotopes as the above-mentioned $^{209}$Bi or $^{76}$Ge we are looking for residential distances that are far greater than the size of our Big Bang universe. That is where they 'actually' are. Does this same logic apply to macroscopic objects like ourselves? Do we 'exist' in some other distant galaxies, or, perhaps, outside 'our' visible universe? These are the issues for some wild (and, perhaps, weird) metaphysical speculations and contemplations that I am leaving at this point as open questions.

Another possible comment here is that the above result (or hypothesis) may open up a new vista on Bell–Bohm-type quantum non-localities and/or the classical interpretation of Newtonian inertia through the action of the 'remote universe'. The spooky partner of Schrödinger's cat may 'in reality' live some $10^{100}$ light years away. Our apparent localization 'here' may, in this context, be an illusion. We, and everything else around us, may (in this picture, at least) be delocalized entities. And, with specific regard to isotopes, we can also note the following. Because the lifetimes of (quasi-) stable isotopes are exponentially different on supercosmic timescales, variations in contributions to QSSs from remote parts of the (mega-) universe (a vastly different $R$ for different isotopes) may drastically enhance the informational aspects of isotopic effects, making isotopic diversity work as the pattern-forming connector between microscopic and ultra-cosmic mega scales.

# References

Aczel A D 2000 *The Mystery of the Aleph: Mathematics, the Kabbalach, and the Search for Infinity* (New York: Four Walls Eight Windows)

Albert D Z 1992 *Quantum Mechanics and Experience* (Cambridge, MA: Harvard University Press)

Arseneva-Geil A N, Berezin A A and Melnikova E V 1976 Photoelectric emission of beta-rhombohedral boron *Fizika Tverdogo Tela* **17** 2448–9 (in Russian)

Arseneva-Geil A N, Berezin A A and Melnikova E V 1976 Photoelectric emission of beta-rhombohedral boron *Sov. Phys.—Solid State* **17** 1624 (Engl. transl)

Bastian G, Vogelsang A and Schiffmann C 2010 Isotopic superlattices for perfect phonon reflection *J. Electron. Mat.* **39** 1769–71

Baz A I, Zeldovich Ya B and Perelomov A M 1966 *Rassejanie, Reakzii I Raspady v Nerelativistskoj Kvantovoj Mechanike* (Moscow: Nauka) (in Russian)

Baz A I, Zeldovich Ya B and Perelomov A M 1966 *Scattering, Reactions and Decays in Non-relativistic Quantum Mechanics* (Jerusalem: Israel Program for Scientific Translation) (Engl. transl.)

Berezin A A 1969 Theory of the Auger effect in a system of two crystal defects *Fizika Tverdogo Tela* **11** 1587–90 (in Russian)

Berezin A A 1969 Theory of the Auger effect in a system of two crystal defects *Sov. Phys.—Solid State* **11** 1285–7 (Engl. transl.)

Berezin A A and Kirii V B 1969 Delta-function potential approximation in the theory of negatively charged electronic color centers *Fizika Tverdogo Tela* **11** 2118–21 (in Russian)

Berezin A A and Kirii V B 1970 Delta-function potential approximation in the theory of negatively charged electronic color centers *Sov. Phys.—Solid State* **11** 1709–11 (Engl. transl.)

Berezin A A 1970 Theory of positron annihilation at F-centers *Fizika Tverdogo Tela* **12** 3315–7 (in Russian)

Berezin A A 1971 Theory of positron annihilation at F-centers *Sov. Phys.—Solid State* **12** 2684–5

Berezin A A, Zaitsev V K, Kazanin M M and Tkalenko E N 1972 Frenkel–Poole effect in (beta)-rhombohedral boron *Fizika Tverdogo Tela* **14** 2813–5 (in Russian)

Berezin A A, Zaitsev V K, Kazanin M M and Tkalenko E N 1972 Frenkel–Poole effect in (beta)-rhombohedral boron *Sov. Phys.—Solid State* **14** 2445–6 (Engl. transl.)

Berezin A A, Golikova O A and Zaitsev V K 1973 Nature of hopping conduction in (beta)-rhombohedral boron *Fizika Tverdogo Tela* **15** 1856–9 (in Russian)

Berezin A A, Golikova O A and Zaitsev V K 1973 Nature of hopping conduction in (beta)-rhombohedral boron *Sov. Phys.—Solid State* **15** 1237–9 (Engl. transl.)

Berezin A A 1973 Theory of the polaron effect in boron *Fizika Tverdogo Tela* **15** 1937–11939 (in Russian)

Berezin A A 1973 Theory of the polaron effect in boron *Sov. Phys.—Solid State* **15** 1298–9 (Engl. transl.)

Berezin A A 1976 Theory of positron localization at negatively charged centers in ionic crystals *Fizika Tverdogo Tela* **18** 858–9 (in Russian)

Berezin A A 1976 Theory of positron localization at negatively charged centers in ionic crystals *Sov. Phys.—Solid State* **18** 493–4 (Engl. transl.)

Berezin A A 1977a Possibility of the Auger effect as a result of positron capture by a multicenter crystal defect *Fizika Tverdogo Tela* **19** 1480–2 (in Russian)

Berezin A A 1977a Possibility of the Auger effect as a result of positron capture by a multicenter crystal defect *Sov. Phys.—Solid State* **19** 864–5 (Engl. transl.)

Berezin A A 1977b Mechanism of autoionization—tunnel relaxation of electron excitation in crystals with oppositely charged defect centers *Fizika Tverdogo Tela* **19** 3372–5 (in Russian)

Berezin A A 1977b Mechanism of autoionization—tunnel relaxation of electron excitation in crystals with oppositely charged defect centers *Sov. Phys.—Solid State* **19** 1969–70 (Engl. transl.)

Berezin A A 1978 Radiation-tunnel relaxation of electronic excitation in alkali-halide crystals with F- and alpha-centers: theory *Optika i Spektroskopya* **44** 261–3 (in Russian)

Berezin A A 1978 Radiation-tunnel relaxation of electronic excitation in alkali-halide crystals with F- and alpha-centers: theory *Opt. Spectrosc. (USSR)* **44** 151–3 (Engl. transl.)

Berezin A A 1979a On the long-lived excited states and the spontaneous tunnel transitions in some positively charged color center systems in alkali halide crystals *Phys. Lett.* **72A** 48–50

Berezin A A 1979b Positron trapping by the negatively charged f-aggregate color centers and the binding energy of the Fe+, Me+ and Re+ centers in alkali-halide crystals *J. Phys. C: Solid State Phys.* **12** L363–6

Berezin A A 1980a Radiative tunnel transitions in some negatively charged colour center systems in alkali halide crystals *J. Phys. C: Solid State Phys.* **13** L103–6

Berezin A A 1980b Radiative tunnel transitions in the hopping conduction in doped semiconductors in a strong electric field *J. Phys. C: Solid State Phys.* **13** L947–9

Berezin A A 1981a On the hopping conduction mechanism of amorphous semiconductors in a strong electric field *Phys. Lett.* **86A** 480–2

Berezin A A 1981b On the theory of the hopping conduction in beta-rhombohedral boron in a strong electric field *J. Less-Common Met.* **82** 143–8 (To the best of my knowledge, this was the first published paper to propose that the isotopic disorder between stable isotopes of boron, $^{10}$B and $^{11}$B may cause electron localization in a complex crystal structure of beta-boron)

Berezin A A 1982a Anderson localization induced by an isotopic disorder *Lett. Nuovo Cimento* **34** 93–6

Berezin A A 1982b On the Anderson transition in electronic color centers systems in alkali halide crystals *Z. Nat.forsch.* A **37** 613–4

Berezin A A 1983a Spontaneous tunnel transitions induced by redistribution of trapped electrons over impurity centers *Z. Nat.forsch.* A **38** 959–62

Berezin A A 1983b Resonance energy transfer in activationless hopping conductivity *Phys. Rev. Lett.* **50** 1520–23

Berezin A A 1983c Radiative unstabilities in some triple electronic colour centre systems in ionic crystals *Phys. Lett.* **95A** 266–8

Berezin A A 1983d Energy transfer processes in non-ohmic hopping conductivity in a strong electric field *Phys. Lett.* A **97** 105–7

Berezin A A 1984a Excited states deactivating exclusively through correlated electron tunneling *J. Chem. Phys.* **81** 851–4

Berezin A A 1984b Instabilities against correlated two-electron tunnel transitions in impurity systems *Solid State Comm.* **49** 87–9

Berezin A A 1984c Two-electron-one-photon transitions in polycenter systems *Chem. Phys. Lett.* **104** 226–8

Berezin A A 1984d Double tunnel jumps in activationless hopping conductivity *J. Phys. C: Solid State Phys.* **17** L393–7

Berezin A A 1984e An isotopic disorder as a possible cause of the intrinsic electronic localization in some materials with narrow electronic bands *J. Chem. Phys.* **80** 1241–5

Berezin A A 1984f Isotopic disorders as a limiting factor for the mobility of charge carriers *Chem. Phys. Lett.* **110** 385–7

Berezin A A 1984g Isotopic biology *Nuovo Cimento* D **3** 914–6 (Discussion of the possibility of alternative biology based on isotopic combinations.)

Berezin A A 1984h Information storage based on isotopic combinations *Specul. Sci. Technol.* **7** 317–9

Berezin A A 1984i Hopping-type electronic processes in pre-breakdown electrical fields in insulators with defect centers *IEEE Trans. Electr. Insul.* **19** 183–5

Berezin A A and Jamroz E J 1984 Collectively deactivating excited configurations of impurity systems in electric field—electroluminescence and switching *J. Lumin.* **31/32** 188–90

Berezin A A 1986a Two- and three-dimensional Kronig–Penney model with delta-function-potential wells of zero binding energy *Phys. Rev.* B **33** 2122–4

Berezin A A 1986b Quasi-localization and electronic transport in boron and boron-rich borides *International Conference on the Physics and Chemistry of Boron and Boron-Rich Solids (Albuquerque, NM, 1985)* vol 140 (New York: AIP) pp 224–33

Berezin A A 1987a On the possibility of isotope ordering and isotopic superlattices *J. Phys. C: Solid State Phys.* **20** L219–21

Berezin A A 1987b Localized states associated with isotopic islands' *J. Phys. Chem. Solids* **48** 853–5

Berezin A A 1987c Isotopic jets as a perfect random number generator *Int. J. Electron.* **63** 673–5

Berezin A A 1987d Localized electron levels in isotopically ordered crystals (isotopic quantum wells) *Phys. Status Solidi* B **144** 727–32

Berezin A A 1988a Isotopic superlattices and isotopically ordered structures *Solid State Comm.* **65** 819–21

Berezin A A 1988b Total internal reflection on isotopic interface: a case for isotopic fiberoptics *J. Opt. Soc. Am.* B **5** 728–9

Berezin A A 1988c Light bending and light confinement at isotopic boundaries—possibility of isotopic fiber optics *Thin Solid Films* **158** L37–8

Berezin A A 1988d Isotopic ordering and isotopic correlations as a possible new tool for geosciences *Chem. Geol. (Isot. Geosci. Sect.)* **72** 197–8

Berezin A A 1988e Isotopic randomness—some fundamental and applied aspects *Phys. Essays* **1** 133–7

Berezin A A and Ibrahim A M 1988 Effects of the diversity of stable isotopes on properties of materials *Mat Chem. Phys.* **19** 407–30

Berezin A A, Chang J S and Ibrahim A M 1988 Isotopically ordered solids—possible electronic and optical applications *Chemtronics* **3** 116–9

Berezin A A 1989a Isotopic engineering (perspectives) *J. Phys. Chem. Solids* **50** 5–8

Berezin A A 1989b Some effects of positional correlations of stable isotopes *Phys. Lett.* A **138** 447–50

Berezin A A 1990 Isotopic randomness and isotopic ordering in phonon physics *Phonons 89, Proc. 3rd International Conference on Phonon Physics and the 6th International Conference on Phonon Scattering in Condensed Matter (Heidelberg, Federal Republic Germany, 21–25 August 1989)* vol 2 ed S Hunklinger, W Ludwig and G Weiss (Singapore: World Scientific) pp 1211–3

Berezin A A 1991 Isotopic relatives of strange attractors *Phys. Lett.* A **161** 295–300

Berezin A A and Ibrahim A M 1991 Semiempirical model of isotopic shifts of the band gap *Phys. Rev.* B **43** 9259–61

Berezin A A 1992a Isotopicity: implications and applications *Interdiscip. Sci. Rev.* **17** 74–80

Berezin A A 1992b Isotopic lattice symmetry: an unexplored area of research *Struct. Chem.* **3** 169–74

Berezin A A 1993a Isotopic effects and corrosion of materials *Mater. Phys. Chem.* **34** 91–100

Berezin A A 1993b Quantum-mechanical aspects of corrosion dynamics *Trends in Corrosion Research* vol 1 (Kaithamukku: Research Trends, Council of Scientific Integration) pp 67–89

Berezin A A 1994a Isotopes, information and quantum self *Frontier Perspect.* **4** 36–8

Berezin A A 1994b The problem of ultimate reality and meaning in the context of information self-organization and isotopic diversity *Ultimate Real. Meaning* **17** 295–309

Berezin A A 1994c Ultra high dilution effect and isotopic self-organisation *Ultra High Dilution. Physiology and Physics* ed P C Endler and J Schulte (Dordrecht: Kluwer) pp 137–69

Berezin A A 1994d Quantum aspects of self-organization in dynamically random systems *Dusty and Dirty Plasmas, Noise, and Chaos in Space and in the Laboratory* ed H Kikuchi (New York: Plenum) pp 225–40

Berezin A A 1995 Electrification of solid materials *Handbook of Electrostatics* ed J S Chang, A J Kelly and J M Crowley (New York: Marcel Dekker) pp 25–38

Berezin A A 2004a Isotopic engineering as a conceptual framework for courses in microelectronics and quantum informatics *Int. J. Eng. Educ.* **20** 4–12

Berezin A A 2004b Isotopic diversity in natural and engineering design *Design and Nature II, 2nd International Conference on Design and Nature: Comparing Design in Nature with Science and Engineering (Rhodes, Greece, June 2004)* ed M W Collins and C A Brebbia (Southampton: WIT Press) pp 411–9

Berezin A A 2006 Simulation argument in the context of ultimate reality and meaning *Ultimate Real. Meaning* **29** 244–61

Berezin A A 2011 Isotopic engineering in surface science and technology *Surface Effects and Contact Mechanics X, 10th International Conference on Surface Effects and Contact Mechanics (Malta, September 2011)* ed J T M De Hosson and C A Brebbia (Southampton: WIT Press) pp 193–204

Berezin A A 2015 *Isotopicity Paradigm: Isotopic Randomness in the Digital Universe* (Cambridge: Cambridge International Science)

Bloch W G 2008 *The Unimaginable Mathematics of Borges' Library of Babel* (Oxford: Oxford University Press)

Bohm D 1952 A suggested interpretation of quantum theory in terms of hidden variables *Phys. Rev.* **85** 93–109

Bohm D and Hiley B J 1993 *The Undivided Universe: An Ontological Interpretation of Quantum Theory* (New York: Routledge)

Borges J L 1998 *Collected Fictions* (New York: Viking)

Bottger H and Bryksin V V 1985 *Hopping Conduction in Solids* (Berlin: VCH)

Cardona M and Thewalt M L W 2005 Isotope effects on the optical spectra of semiconductors *Rev. Mod. Phys.* **77** 1173–224

Chen X, Chang J S, Berezin A A, Ono S and Teii S 1991 Isotope effects in amorphous silicon thin films produced by Ar-SiH4-D2 plasma chemical vapor deposition method *J. Appl. Phys.* **69** 1678–86

Cranston E D and Gray D G 2006 Formation of cellulose-based electrostatic layer-by-layer films in a magnetic field *Sci. Technol. Adv. Mater.* **7** 319–21

Dawkins S 1976 *The Selfish Gene* (Oxford: Oxford University Press)

Demkov Yu N and Ostrovsky N V 1975 *Metod Potentsialov Nulevogo Radiusa v Atomnoi Fizike* (Leningrad: Leningrad State University Press) (in Russian)

Demkov Yu N and Ostrovsky V N 1988 *Zero-Range Potentials and Their Application in Atomic Physics* (New York: Plenum) (Engl. transl.)

De Sa MS and Berezin A A 1989 The Application of catastrophe theory to corrosion problems *Corros. Sci.* **29** 1141–8

Derbyshire J 2004 *Prime Obsession: Bernhard Riemann and the Greatest Unsolved Problem in Mathematics* (New York: Plume)

Deutsch D 1997 *The Fabric of Reality* (London: Allen Lane)

Eberhart M E 1999 Why things break *Scientific American* (October 1999) pp 66–73

Eberhart M E 2003 *Why Things Break* (New York: Harmony)

Ehrenberg W 1967 Maxwell's demon *Scientific American* (November 1967) **217** 103–10

Feiler A A, Bergstrom L and Rutland M W 2008 Superlubricity using repulsive Van der Waals forces *Langmuir* **24** 2274–6

Feldmann T and Kosloff R 2006 Quantum lubrication: suppression of friction in a first-principles four-stroke heat engine *Phys. Rev.* E **73** 025107

Fleischmann M and Pons S 1989 Electrochemically Induced Nuclear Fusion of Deuterium *J. Electroanal. Chem.* **261** 301–8

Fleischmann M, Pons S and Hawkins M 1989 Electrochemically Induced Nuclear Fusion of Deuterium *J. Electroanal. Chem.* **263** 187–8 (erratum)

Goldman C and Berezin A A 1995 Isotopic fractionation by phonon induced interactions *Phys. Rev.* B **51** 12361–8

Haken H 1978 *Synergetics—Nonequilibrium Phase Transitions and Self-Organization in Physics, Chemistry and Biology* 2nd ed (Berlin: Springer)

Haller E E 1995 Isotopically engineered semiconductors *Appl. Phys. Rev.* **77** 2857–78

Haller E E 2002 Isotopically controlled semiconductors *J. Nucl. Sci. Technol.* **39** 382–5

Haller E E 2005 Isotopically controlled semiconductors *Solid State Comm.* **133** 693–707

Hoffman A and Scherz U 1990 Jahn–Teller effect and zero-phonon line isotope shifts of transition metals in II-VI compounds *J. Cryst. Growth* **101** 385–92

Ibrahim A M and Berezin A A 1992 Synthesis of buried insulating layers in silicon by ion implantation *Mater. Chem. Phys.* **31** 285–300

Ishida T 2002 Isotope effect and isotope separation: a chemist's view *J. Nucl. Sci. Technol.* **39** 407–12

Itoh K M, Haller E E, Hansen W L, Beeman J W, Farmer J M, Rudnev A, Tikhomirov A and Ozhogin V I 1994 Neutron transmutation doping of isotopically engineered Ge *Appl. Phys. Lett.* **64** 2121–3

Itoh K 2009 Diamond nanostructures: isotopes for nanoelectronic devices *Nat. Nanotechnology* **4** 480–1

Itoh K and Watanabe H 2014 Isotope engineering of silicon and diamond for quantum computing and sensing applications *MRS Comm.* **4** 143–57

Jiang X-L and Berezin A A 1998 Channeling effects and nuclear reactions in electrochemical systems *J. New Energy* **3** 84–92

Kashkai A D, Berezin A A and Arseneva-Geil A N 1972a Photo-emission from F-aggregate centers in KCl and KBr crystals *Fizika Tverdogo Tela* **13** 3130–1 (in Russian)

Kashkai A D, Berezin A A and Arseneva-Geil A N 1972a Photo-emission from F-aggregate centers in KCl and KBr crystals *Sov. Phys.—Solid State* **13** 2633–4 (Engl. transl.)

Kashkai A D, Berezin A A, Arseneva-Geil A N and Matveev M S 1972b Photoemission studies of electronic color centers in additively colored KCl and KBr crystals *Phys. Status Solidi* B **54** 113–9

Kervran C L 1972 *Biological Transmutations* (Bristol: Crosby Lockwood)

Klemens P G 1981 Thermal conductivity of pure monoisotopic silicon *Int. J. Thermophys.* **2** 323–30

Kojima T, Nebashi R, Itoh K M and Shiraki Y 2003 Growth and characterization of $^{28}$Si $–^{30}$Si isotope superlattices *Appl. Phys. Lett.* **83** 2318–20

Lamoreaux S K 2009 Quantum force turns repulsive *Nature* **457** 156–7

Landau L D and Lifshitz E M 1963 *Kvantovaya Mekhanica, Nerelyativistskaya Teoriya (Teoreticheskaya Fizika, Tom III), (Quantum Mechanics, Non-relativistic Theory, Theoretical Physics)* vol 3 (Moscow: Fizmatgiz)

Leff H S and Rex A F (ed) 1990 *Maxwell's Demon: Entropy, Information, Computing* (Princeton, NJ: Princeton University Press)

Lloyd S 2002 Computational capacity of the Universes *Phys. Rev. Lett.* **88** 237901

Lloyd S 2006 *Programming the Universe: A Quantum Computer Scientist Takes on the Cosmos* (New York: Knopf)

Mukherjee S *et al* 2015 Phonon engineering in isotopically disordered silicon nanowires *Nano Lett.* **15** 3885–93

Munday J N, Capasso F and Parsegian V A 2009 Measured long-range repulsive Casimir–Lifshitz forces *Nature* **457** 170–3

Nicolis G and Prigogine I 1977 *Self-Organization in Nonequilibrium Systems: From Dissipative Structures to Order Through Fluctuations* (New York: Wiley)

Penrose R 1994 *Shadows of the Mind* (Oxford: Oxford University Press)

Penrose R 1996 On gravity's role in quantum state reduction *Gen. Relativ. Gravit.* **28** 581–600

Plekhanov V G 2004 *Applications of Isotopic Effect in Solids* (Berlin: Springer)

Plichta P 1997 *God's Secret Formula: Deciphering the Riddle of the Universe and the Prime Number Code* (Rockport, MA: Element)

Pui J P and Berezin A A 2001 Mind, matter, and diversity of stable isotopes *J. Sci. Explor.* **15** 223–8

Ribenboim P 1989 *The Book of Prime Number Records* 2nd edn (New York: Springer)

Shklovskii B I and Efros A L 1984 *Electronic Properties of Doped Semiconductors* (Berlin: Springer)

Tegmark M 2014 *Our Mathematical Universe: My Quest for the Ultimate Nature of Reality* (New York: Knopf)

von Neumann J and Wigner E P 1929 Uber merkwurdige diskrete Eigenwerte *Physikalische Zeitschrift* **30** 465–7

Watanabe H, Nebel C E and Shikata S 2009 Isotopic homo junction band engineering from diamond *Science* **324** 1425–8

**IOP** Publishing

Digital Informatics and Isotopic Biology
Self-organization and isotopically diverse systems in physics, biology and technology
**Alexander Berezin**

# Chapter 5

## Isotopicity in biology and in the theory of consciousness

*I regard consciousness as fundamental. I regard matter as derivative from consciousness. We cannot get behind consciousness. Everything that we talk about, everything that we regard as existing, postulates consciousness.*

Max Planck (1858–1947)

*I think it's science and physics are just starting to learn from all these experiments. These experiments have been carried out hundreds and hundreds of times in all sorts of ways that no physicist really questions the end point. I think that these experiments are very clearly telling us that consciousness is limitless and the ultimate reality.*

Robert Lanza (b 1956)

This chapter reviews the ideas and hypotheses that the author sees as the most important line of the discourse in this book. It is a common point that science thrives on hypotheses and this very term has several synonyms, and among these is the often-used word 'speculation'. For some people the latter term has some questionable and suspicious associations. Notions of UFOs, poltergeists, 'reptilians among us' and 'conspiracy theories' may come to mind. And, yes, to some degree such associations are often justified. However, as the history of science (and technology) amply demonstrates, many so-called speculations turned out to be the foundations of many important developments and true breakthroughs. When Alfred Wegener came up with the idea of continental drift it was almost uniformly rejected as a wild speculation—before it was experimentally confirmed and became one of the most fundamental components of modern geophysics and geology (Oreskes 1999). Many more examples of the same kind could easily be quoted here.

To restate again what was said earlier and may be said again later in this book, this author makes no excuses for the fact that many, perhaps the majority, of the ideas and inferences presented here can be categorized as speculations. But, again, who are the judges here? People at all levels of professionalism and achievement often have greatly divergent views on the same subjects. It is sufficient to recall that when Georg Cantor came up with his ideas on the cardinality of infinite sets (alephs), two truly great scientists in their own way—Henri Poincaré and David Hilbert had absolutely opposite reactions. The first (Poincaré) called Cantor's idea insane, while David Hilbert admired Cantor's set theory and said that 'no one shall drive us from the paradise which Cantor has created for us'. And the words of Cantor himself—'in mathematics the art of proposing a question must be held of higher value than solving it'—can probably also be applied to other sciences than pure mathematics, at least in many instances (Dauben 1979).

On this note, this chapter outlines several ideas on the possible role of isotopicity and isotopic randomness in biological, biomedical and cognitive sciences—even if many aspects of these ideas can only be experimentally tested decades from now.

## 5.1 Mechanistic and holistic approaches to consciousness

*We've discovered that the earth isn't flat; that we won't fall off its edges, and our experience as species has changed as a result. Maybe we'll soon find out that the self isn't 'flat' either, and that death is as real and yet as deceptive as the horizon; that we don't fall out of life either.*

Jane Roberts (1929–1984), Seth Speaks

It is not my intention within the format and constraints of this book to give even a brief review of the key ideas and developments regarding consciousness. Or, properly speaking, what we call 'consciousness'. And here the multiplicity of meanings and contexts is truly astonishing. The huge and ever-growing tsunami of literature of all forms shows no sign of receding. On the contrary, all the evidence suggests that this process most likely will continue at an almost exponential rate (Kurzweil 2005).

However, with all due qualifications regarding the conditionality and relativity of all the conceptual demarcations, we can envision two lines of attack on the issue of consciousness. Let me call them (provisionally) the mechanistic and the holistic. Depending on the particular details of the discourse, these two trends (or, one might say, two camps) can be considered as being opposed or complementary to each other. Sometimes it is the personal preferences and sympathies of a particular author that determine which side (mechanistic of holistic) the discussion tends to tilt towards. This author prefers a blended approach to consciousness that combines mechanistic (reductionist) and holistic (unifying) ends. Much of what is written on the quantum nature of consciousness fuses both approaches to various degrees of synthesis and/or eclecticism. It is my view that this is a healthy situation because it fosters ongoing debate and the throwing out of new and fresh ideas in an area where, most likely, there is not much chance of a 'final truth' emerging.

## 5.2 The observer effect in quantum physics

*Science predicts that many different kinds of universe will be spontaneously created out of nothing. It is a matter of chance which we are in.*

Stephen Hawking (b 1942)

Much of the discourse on the nature and mechanisms of consciousness revolves around the so-called 'observer effect' in quantum physics (Penrose, Deutsch, Zohar and many others). The idea that observation creates reality can be traced back to much earlier philosophical and metaphysical contemplations (in kaons such as 'Is there a Moon if nobody looks at it?'), but the observer effect was only recently put onto a solid physical foundation through delayed choice and quantum entanglement experiments. Interestingly enough, this line of questioning is strongly intermingled with the idea of parallel Universes and is becoming popular in the cognitive sciences and psychology (e.g. Robert Lanza and others, see, e.g. Lanza and Berman 2009).

The recent popularity of these ideas to a large degree comes from their close connection to the human realm and our existential milieu. Such ideas as 'we never die' certainly to be appear exciting and uplifting. Robert Lanza and others assert such a notion on the basis of the theory of parallel Universes. The latter is also known as the many world interpretation of quantum physics, which originated with Hugh Everett and was later developed further by many other authors.

Without going into too much detail regarding that here (the literature and web sources are abundant), suffice it to say that in a multitude (infinity) of parallel Universes we exist in infinitely many copies and in all sorts of variations. So, if we die in this particular Universe (for example, a proverbial brick falls on our head), our identity is immediately transported to another Universe where the brick avoided us (or fell on somebody else's head). And no matter how far-fetched and unbelievable this may appear, the arguments are quite extensive, based on substantial quantum physics (say, double slit and delayed choice experiments) and are gaining momentum in serious (and even mainstream) scientific discourse.

## 5.3 The holomovement of David Bohm and universal entanglement

*Science is a way of thinking much more than it is a body of knowledge.*

Carl Sagan (1934–1996)

In much of the literature on the observer effect and universal connectedness the ideas of David Bohm's *holomovement* take the central seat. And rightly so, because among the elite group of top international quantum physicists David Bohm (1917–92) spelled out the ideas of universal quantum connectedness ('oneness') in the strongest and most penetrative way. This is why many people, in particular those with a 'spiritual' and New Age orientation, see Bohm as the godfather of these ideas (e.g. Kafatos and Nadeau 1990, Siler 1990, Zohar 1990, Arnold 1992, Talbot, 1992,

McTaggart 2003 and many other popular authors who write at the interface of physics and holistic viewpoints).

Quantum physics as it is understood today is a theoretical construct that places a strong emphasis on non-locality (Penrose 1994, Deutsch 1997, Grib and Rodrigues 1999, Lloyd 2006). One of the most interesting effects of quantum non-locality is quantum entanglement, which basically posits that spatially separated objects still retain their inner connection. Even when two quantum particles (say, electrons or photons) are located at opposite edges of the galaxy, they still remain intimately connected (they 'know what the other partner is doing', so to speak).

This phenomenon of quantum entanglement has recently begun to focus on a variety of areas that range from quantum informatics through to nanotechnology and the biomedical sciences. The surfaces and contacts of microelectromechanical systems (MEMS)-scale devices is another exploratory area (Berezin 2011). In this regard, isotopic purification and isotopic engineering could be instrumental for strengthening the quantum coherence and connectivity of nanoscale structures and devices.

I will not discuss every aspect and controversy of non-locality in quantum physics here as they are well covered in numerous easily available sources in the literature. Suffice it to mention such fascinating facets as instant (perhaps, super-luminal) connections between distant objects (e.g. two atoms in different galaxies may still be instantaneously connected), teleportation, Bell's theorem and speculations about 'backward causality' (the future may affect the past, or, equivalently, 'time loops' may exist). Here I make/ask some (tentative, of course) suggestions/questions about possible links between quantum non-locality and various aspects of isotopicity and the inferences that can be drawn from them.

In order to more vividly perceive the idea of quantum entanglement, we can use a universally known classical effect of magnetism. We, as children and adults alike, love to play with magnets. When we take two magnets and try to move them together with the same poles facing each other we feel that they resist. It is like there is some elastic spring between them that we are trying to compress. But there is nothing (*nothing*) visible between them. And this will certainly work the same way in a vacuum when there is no air around. Of course, thanks to Maxwell and many others, there is a formal theory of magnetism which talks about the energy of the magnetic field and provides a good mathematical description of magnetic actions and all of that. And yet, it does not much help our direct feelings and senses to comprehend this phenomenon. Such longstanding issues as the 'true' nature of photons and quantum vacuum remain enigmatic and speculation regarding them often stands at the borderline between physics and metaphysics (e.g. Zajonc 2003). As for magnets, we still remain open-mouthed and puzzled by this enigma: how do these magnets 'know' each other and how do they communicate?

While these effects had been discovered and described long before the advent of quantum physics (or indeed, *any* physics for that matter), they can be seen as down-to-earth illustrations of non-local effects. Magnetism as a precursor of (quantum) non-locality and the very idea of a field (be it magnetic, electrostatic or gravitational) has an inherent notion of non-locality built into it. From these analogies we can move to quantum non-localities in a proper sense.

Some examples of non-locality (albeit in some metaphorical sense) can be found even in the area of our human perceptions. A popular illustration of this kind of non-physical non-locality is a rainbow. We see a rainbow in the sky, as if it is 'up there'. And we perceive it as a whole, as some kind of connected object. But in reality, a rainbow is the result of some spectral perception in our eyes; if we fly up into the sky (say, by helicopter) to 'catch' a rainbow, we will find nothing except moist air. The rainbow 'exists' as a result of the play of waves of light in our eyes and the processing of this by our brain/mind. It is a bit like those objects that we see in a mirror (like our own reflections), they do not actually exist in the mirror, but only in our perceptual system.

## 5.4 Biological implications of isotopic diversity

*The scientist is not a person who gives the right answers, he is one who asks the right questions.*
Claude Levi Strauss (1908–2009), French anthropologist

Almost simultaneously with my work on isotopically induced Anderson localization, I was thinking about the possibility that combinations of isotopes could carry biological information in a similar (and/or complementary) fashion to regular hetero-atomic chemistry. Unquestionably, in the eyes of many such an idea appeared highly speculative. However, in 1984 the internationally known journal *Naturwissenschaften* (Springer) published my short (one-page) article with the somewhat provocative title 'Can life be based on a single chemical substance?' (Berezin 1984e). In it I suggested (hypothetically) that the information potential of isotopic combinations could provide a possibility for the evolution of 'isotopic life' as an alternative (or complementary) realm to biology based on a hetero-atomic chemistry.

Any substance which has a natural isotopic diversity (e.g. water) can, in principle, form the basis for an independent 'isotopic biology', hence the hypothetical ocean of water that, like the one suggested by Stanislaw Lem in his novel *Solaris* (Lem 1981), could indeed be 'alive' and, perhaps, even conscious. Such ideas as 'the Sun is alive' (Sams 2009) also fall into the same fold.

In the next few sections I present an unfolding of this idea along several directions. One of the questions that I tackle is: what can isotopes do in biology that common chemistry can not? By 'common chemistry' I mean the chemistry of the elements of the traditional periodic table without any consideration of the isotopic diversity of elements, as if all elements just had a single isotope (a chemistry that, for the most part at least, largely ignores isotopic diversity and the (generally subtle) effects of isotopicity). In this way 'isotopic chemistry' runs over and above common chemistry as a kind of 'shadow chemistry' embedded into 'normal' chemistry.

One of the primary founders of quantum physics, Erwin Schrödinger (1887–1961), was also among the first to address quantum aspects of biological life

(Schrödinger 1945). His main thesis was that quantum mechanical indeterminism can generally be seen as a potential source of unpredictability and variability in living beings. In terms of the behavior of a given organism under a fixed set of fully (classically) described conditions, quantum mechanical indeterminism was traditionally pinpointed as the most obvious microscopic generator of the individual behavioral patterns of organisms at practically all levels of complexity.

Without denying this generic role of 'regular' quantum mechanical indeterminism, one can take an independent look at the very phenomenon of isotopicity and see isotopic diversity in Nature as an alternative powerful source of randomness, irreproducibility and individuality in living matter (Berezin 1987c). In this context, isotopicity could act as a freedom-enhancing factor through random and unrepeatable positional combinations of various isotopes of the same chemical element within functionally critical biological structures.

As was mentioned earlier, the majority of biologically important elements (H, C, N, O) exist in Nature as a mixture of two (or more) stable isotopes. And as was also said above, it is well known that the physical characteristics of substances with different isotopic content are generally close, but not exactly the same. Although carbon is mostly $^{12}$C, oxygen is mostly $^{16}$O and nitrogen is mostly $^{14}$N, the quantity of 'minority' stable isotopes is by no means negligible (e.g. 1.1% of $^{13}$C, 0.2% of $^{18}$O, 0.36% of $^{15}$N, etc). Even a seemingly small relative abundance of a minority isotope—say, 0.01% (one per 10 000)—translates into huge absolute concentrations like $10^{17}$–$10^{18}$ cm$^{-3}$.

Thus far, isotopic diversity is unavoidable even in a very small segment of a genetically important molecular structure. One should appreciate the fact that although isotopic variations of bond lengths and strengths, atomic mobility, and other structural and dynamical parameters are very small, they are present virtually everywhere and cumulatively the influence of such isotopic diversity could be significant. Small but omnipresent isotopic fluctuations of bond lengths, reaction rates, etc, could provide an additional degree of freedom for individual operating characteristics, as well as for evolutionary processes at large.

Isotopic randomness leads to the situation that virtually no two nominally identical complex molecules (e.g. 'Xerox copies' of DNA fragments) are, strictly speaking, the same. Natural isotopes of H, C, N, O, etc are randomly distributed among the corresponding atomic positions. Let us consider a large molecule which is chemically fully specified, e.g. a fixed fragment of DNA. The number of different combinations of isotopes in the molecule, which chemically remains nominally the same, is practically uncountable.

For example, for a DNA molecule with 100 000 atoms, this number is roughly of the order of $10^{30000}$ (10 to the power of 30 000), and this number is vastly greater than the total number of atoms in the known (Big Bang) Universe (the latter is 'only' about, say, $10^{90}$ or $10^{100}$, and estimates such as $10^{90}$ or $10^{100}$ do not make much difference to our human perception). These estimates belongs to a special class of super-large numbers, or tower exponents—a class of numbers to which the famous Skewes number belongs (Skewes (1933), (1955), Knuth (1976), Berezin (1987d)). The very fact of isotopic uniqueness (and, therefore, the virtual irreproducibility) of any complex molecular structure could serve as an ample resource of unpredictability at the level of single atoms. For all practical purposes (FAPP), the number of possible isotopic combinations should be considered as virtually infinite.

My model of isotopic biology suggests the possibility of using isotopic combinations for an alternative mechanism of storage of genetic information. In this regard we can take another look at the possible role of isotopic permutations in the key informational molecules in living systems. As already pointed out, for any specified molecular structure the number of possible isotopic permutations is virtually unlimited. This isotopic degree of freedom could be of special importance for the molecular mechanisms of mental processes, memory, creativity, etc. It is generally accepted that single-site mutations occur at the level of single atoms and, consequently, the effectiveness of the above isotopic option seems to be a physical possibility. In short, DNA and other key biological structures could perform differently depending on their individual and irreproducible isotopic signatures.

Some of these ideas could perhaps be tested experimentally in a relatively straightforward way. Some of the available experimental possibilities are briefly discussed below. Most of the ideas in this area are based on a deliberate shift of the isotopic abundances of stable isotopes in living organisms. This may allow us to investigate some physiological and behavioral correlations with isotopic content and composition. I should stress that the subject of isotopic replacements in biology should not be confused with studies on radioactive and stable trace isotopes, which constitute one of the pillars of nuclear medicine. However, the latter topic is outside the primary discussion line of this book.

The physical characteristics of isotopically pure substances are generally quite close, but not exactly the same. We live in an environment of fixed (more-or-less) isotopic ratios. One may ask what will happen to a living organism if we intentionally shift naturally occurring isotopic ratios. It would be interesting to know, for instance, if such a change could affect the rate of metabolism, aging processes, adaptability, etc.

Several authors have discussed the unexpected roles which isotopes can play in biology. Mann and Primakoff (Mann and Primakoff 1981) suggested that biological chirality (asymmetry between the left and right) might originate from the beta-decay of isotopes $^{14}$C and $^{40}$K (K = potassium). Likewise, I suggested in 1984 (Berezin 1984i) that the decay of $^{14}$C atoms can be considered as a possible contributing factor to carcinogenesis and genetic mutations, while Keswani (Keswani 1986) discussed the random disintegrations of $^{14}$C atoms as a likely source of spontaneous and unpredictable triggerings of brain processes such as sudden memory flashes.

The occurrence of the majority of chemical elements in two or more stable isotope forms is one of the primary facts of nature. Therefore, even from a general point of view it seems unlikely that the diversity of stable isotopes could remain totally irrelevant to Earth's biology. It is known that microorganisms can shift the isotopic abundances of heavier elements, e.g. $^{32}$S and $^{34}$S (Thode 1980). These shifts, however, are due to tiny differences in reaction rates leading to preference in isotope incorporation and fractionation. The observed shifts usually do not exceed 1% in the relative change of isotopic proportions in comparison with standard abundances. At the same time, reactions that are sensitive to nuclear magnetic moments can lead to much greater segregation efficiency.

While the mass difference between stable isotopes is at best one of a few per cent (except, of course, for the H–D, hydrogen and deuterium, pair), the difference in the nuclear magnetic moment could be much more drastic. For instance, $^{12}C$ has a zero nuclear magnetic moment, while $^{13}C$ has a non-zero value and, consequently, responds quite differently to magnetically linked perturbations (e.g. oscillatory microscale and mesoscale electromagnetic fields due to noise fluctuations or outside influences, etc). Using reactions involving magnetic (hyperfine) interactions, a single-stage segregation efficiency of about 50% has been obtained for $^{13}C$ and $^{12}C$ (Epling and Florio 1981).

Numerous studies on heavy water enrichment in living organisms have been undertaken since the discovery of deuterium in 1932. In his review, Katz (Katz 1960) gives interesting examples of deuteration effects on mice and rats, showing that these animals cannot survive the replacement of more than about one-third of their body water with heavy water $D_2O$. Most of the changes induced by deuteration (e.g. loss of reproductive capacity) are reversible upon the replacement of deuterium by ordinary hydrogen. Katz (1960) also mentioned the possibility of isotopic replacements for C, O, N and S.

Almost certainly there were other subsequent studies on isotopic replacements in biology, biological isotopic enrichment and other related effects of which I may not be aware. While I make some comments to that effect below, it is not my purpose to provide a comprehensive or updated review of any of this work in this book. My book aims to put forward thought-provoking and mind-stimulating ideas, not to lay any claims to priority or originality. Neither in this book nor in any of my prior papers, do I proclaim that I was/am the first to say such and such a thing. Consequently, I am in advance dismissing any possible charges that somebody proposed this-or-that idea before I published it. If this is the case for any idea, I will certainly recognize and respect the priority of such a person or people. All I can say is that up to the time of writing (2016), no one has ever informed me (by any channels or means) of their priority regarding any of the isotopicity issues discussed in my previously published papers.

## 5.5 The concept of isotopic biology

While the fact that most (about 2/3) of all chemical elements have two or more stable isotopes is generally known, the implications of this for biology have only been discussed rather episodically. However, because of this fact alone, it is natural to ask if the isotopic diversity of key biological elements (H, O, C and N) plays some non-trivial role in bio-evolution up to the level of consciousness (Berezin (1984e), (1984g), (1986b), (1987a), (1987c), (1988e), (1990b), (1992a), (1994a), (1994b), (1994c), (1996), (1998), (2002), (2004), (2015), Pui and Berezin (2001)). In particular, in some of these papers I put forward some ideas about a possible role for the diversity of stable isotopes in brain function. My hypothetical suggestion was that the information processing associated with consciousness could be assisted by the inclusion of isotopic effects in neural information processing at the microscopic level.

When considering the development of these ideas regarding a possible role for isotopic effects in consciousness (my papers from the 1980s–2000s referenced above), it is natural to put it into a more general context. In other words, it is tempting to explore possible connections between isotopic information dynamics and more universal levels of information. In this vein, I proposed that the isotopic degree of freedom (isotopic randomness) could perhaps serve as an 'informational connector' to the fundamental resource of pattern formation which is ultimately available in a form of 'absolute' mathematical relationships. The latter is sometimes referred to as the ideal Platonic world (IPW) of forms and numbers. A precursor to the latter connection is implicit in the mathematical theory of infinite sets (the so-called hierarchy of aleph sets) proposed by Georg Cantor over a century ago (Dauben (1977), (1979), Aczel (2000)).

The availability of chemical elements on Earth has spawned a nearly unlimited variety of structures and organisms through variation of their chemical composition. This resulted in an enormously rich diversity of biological life on our planet. Reflecting on this, one can say that it appears almost certain that by finding some biological role for more-or-less all the chemical elements (including most microelements) Nature optimizes the resources of chemical diversification available to it. In other words, it looks like Nature is smart enough to take every possible advantage from the chemical diversity of elements, including, perhaps, even some radioactive elements.

In this regard we can ask the same question about isotopic diversity. Is, in fact, a similar possibility likely to arise for the isotopic diversity of elements? It seems highly improbable that clever Nature could 'overlook' the additional level of informational diversification that is available through the isotopic degree of freedom and isotopic randomness.

Different stable isotopes of the same chemical element are distinguishable entities. They are distinguishable by mass, nuclear magnetic moment, the position of energy levels and spectral lines, and other quantum properties. Metaphorically, we can talk about isotopic individuality at the atomic level. Due to such distinguishability, the string of different isotopes of the same element is mathematically homomorphic (equivalent) to an information-carrying digital string. For example, strings of carbon isotopes like $^{12}C$ $–^{13}C–^{13}C–^{12}C–^{12}C–^{12}C–^{13}C$ (etc) are equivalent (in terms of information) to a binary string 1001110 ($^{12}C = 1$, $^{13}C = 0$). Elements of more than two stable isotopes can form ternary (etc) strings or three-dimensional information-carrying clusters. Adding the possibility of incorporating trace radioactive isotopes (e.g. $^{14}C$) to this brings an additional enhancement of informational and dynamical options for atomic-scale biological and quasi-biological systems.

For example, it may be possible that during photosynthesis in plants (and depending on the species of the plant), the carbon obtained from $CO_2$ (which in air consists of $^{12}C$ and $^{13}C$ with a minute fraction of radioactive $^{14}C$) may show some isotopic selectivity. In other words, one of these isotopes can be incorporated preferentially. The degree of such fractionation (isotopic selectivity) may be small and may vary from plant to plant, but I see no clear reason to rule out such a possibility up-front. It could even be of the opposite sign for some plants (some plants may 'like' $^{12}C$ more, while others may find $^{13}C$ tastier).

In such a scenario, in the production of energy in the form of adenosine triphosphate (ATP) the various carbon isotopes may be selectively placed so that they are propagated throughout the series of reactions in that same position. This conservation of isotopic structure may persist in spite of the fact that the catalysis of enzymes changes the carbon skeletal structure of the intermediate molecules (Pui and Berezin 2001). Perhaps a key point here is that the effects of thermodynamic randomization may not completely overshadow (or cancel out) the said selectivity (no matter how small the degree of such selectivity is) among different isotopes which (macroscopically and biologically) can be distinguished by their subatomic properties (such as spins, magnetic moments, nucleus size—which differ for various isotopes of the same element).

Elementary combinatorial analysis leads to an enormously large number of possible isotopic permutations within chemically fixed structures. For example, a segment of a DNA molecule with one million carbon atoms has about 10 000 randomly distributed $^{13}C$ atoms. The number of isotopically distinguished distributions (the number of possible placements of the 10 000 atoms among 1 000 000 sites) is about $10^{24000}$. This is a far greater number than the number of atoms in the Universe, which is 'only' estimated to be between $10^{80}$ and $10^{100}$ (the uncertainty may be due to the (currently obscure) 'dark matter', whether it exists or not). If we include the spatial arrangements that can be produced by point substitutions in other stable isotopes, such as $^{16}O$ by $^{17}O$ and $^{18}O$, or $^{14}N$ by $^{15}N$, etc, the possibilities for information transfer and information diversification carried parallel to 'macros' information (such as the genetic transcription of codons or chromosomal crossover) increase even further (tower exponentially).

## 5.6 Isotopic biology and subtle genetic messages

*No stones can ever fall from the sky, because there are no stones in the sky!*
Antoine Lavoisier (1743–1794), the father of modern chemistry,
speaking when he was appointed to chair a French Academy Committee
to study reports of meteorites.

Turning to consciousness-related mental information, one first of all has to address the problem of memory size (Pui and Berezin 2001). As various estimates have it, we require storage facilities for some $10^8$ distinct memory patterns. The human brain maximally possesses approximately $10^{12}$ neurons. The current prevailing model of brain function is an elaboration of the original Cajal hypothesis. For the coordination of movements, actions, speech, decisions, emotions and inner thoughts (all of which we can also remember) with our memories, we would require about $10^{20}$ neurons. According to the Cajal hypothesis, neurotransmitter molecules such as serotonin and dopamine, as well as proteins, effect long-term changes in neuron terminals. These changes are thought to underlie long-term memory. However, there is an accumulating body of evidence to suggest that the information is also conveyed diffusely in cerebral fluid ('volume transmission'), rather than by electrical transmission alone.

We suggested earlier (Pui and Berezin 2001) that the deficiency in the information processing and storage capacity stated above may be compensated by isotopicity in the following way. Without the constraint of wiring, the signals transferred by volume transmission can diffuse over a large brain surface area, interacting with deep brain structures, with several stops along the way for information replication and exchanges. To give a general overview, isotopes could be involved in the receptor-specific actions of a variety of the neurotransmitters involved in volume transmission. It was shown that the molecules formed in succession in enzymatically catalyzed pathways conserve $^{13}C$ in precisely the same position from the beginning to the end of the pathway. It was also pointed out that the positions of $^{13}C$ atoms are non-randomized, and are predetermined by (so far) unknown mechanisms.

While identical isotopes are indistinguishable quantum particles (like electrons), different isotopes are distinguishable in the sense of classical physics. For example, the classical distinguishability of $^{12}C$ and $^{13}C$ isotopes in DNA chains can drastically affect the level of quantum coherence for holographic-type memory storage. Another effect may arise from nuclear spin–spin interactions between $^{13}C$ isotopes ($^{12}C$ has zero nuclear spin). The fact that different isotopes obey different quantum statistics (e.g. $^{12}C$ = bosons and $^{13}C$ = fermions) may also reflect on processes of information transfer at the subatomic level.

From this perspective, nuclear spins and quantum coherence do not seem to mean much, until we observe that various kinds of organisms have been shown to differentiate between isotopes, and to form isotopic pools. Furthermore, it has been shown that metabolic processes and structures are altered, stressed, or become more efficient depending on the type of stable isotopes used. Since isotopes differ in their mass and nuclear spins, and given that organisms demonstrate distinguishability between stable isotopes, we can hypothesize that isotopes add to the beneficial aspects of consciousness by: (1) coping with the surplus of information in the environment and maintaining information coherence within our brains and/or (2) enabling anticipation through the coordination of information.

The possibility of anticipating events (which is often claimed by humans) brings us to the subject of consciousness and attention with respect to isotopic diversity. The ongoing debate about the nature of time perception in humans involves consciousness regardless of the position one wishes to take (whether our sense of time is merely a by-product of our sensory systems or not). To propose a substrate, tangible substance for the characteristics of consciousness, we must include the attempts of our consciousness to extrapolate into infinity, and its universal durability. Not only are isotopes the 'basic stuff' of the Universe, they also have multi-level participation in every organic system. Isotopes are necessarily a part of molecules, organs, organisms and entire communities of living systems, and we can speculate that they play a decisive role in connecting all these biological structures and functions to the fundamental metaphysical substrate of the Universe (the ideal Platonic world, IPW), which serves as a resource for all pattern-forming activities.

The concept of the IPW, where integers are immortal, infinite and immutable, defies the 'physical' order, which may be created or destroyed in any particular moment. The immutability of numbers provides us with as close a view of infinity as we humans can have. The 'weight', or 'pressure', of invariable integers may manifest itself in time-related matters, such as perceptions and thoughts. Combining the previously discussed local and cosmic aspects of isotopic information, isotopes are numerically placed in chemical substances, forming islands of isotopes, and also remain basic components of the complex Universe.

For the infinite set of integers to be a candidate for an organically based ageless consciousness of individuals, they require some kind of a presence in the physical world. Because isotopic configurations in biological systems can be mapped into the sequences of integer numbers (e.g. through their instantaneous coordinates), isotopic quantum dynamics, such as tunneling effects (Berezin 1992a) could act as a bridge towards the process of this embodiment. Such an embodiment (metaphysical patterns of numbers become a physical reality) can be interpreted as a transcendence of the infinite 'innocence' (lack of 'real' experience) that numbers have as entities of the timeless Platonic world.

Neurons have constantly varying inputs, associations and outputs in normal conditions. We can consider this as a finite analogue of the lowest infinite set, aleph zero (countable set). But neurons are only one part of consciousness, which relies on the varying input and output for continued existence. While we are able to determine some of the principles of consciousness, we face monumental difficulties in determining or examining the whole. Just as there are sub-conscious levels of which we are seemingly unaware, likewise, for the first countable set of integers, we have only a vista of its whole range, and examination of this set does not allow for a simultaneous examination of the individual particulars of each integer. For example, two is an integer which belongs to the lowest infinite set (aleph zero), but the (irrational) square root of two is a real (not rational) number at a higher level of infinity (real numbers constitute the next aleph, aleph one set). Thus, the concept of the square root does not belong to the set of integers, no matter how far into infinity we can view the integer (countable) set. Consciousness exists within the first set of infinity, the set of neuronal activity, and in a higher set of infinity.

Although isotopes are finite, they may possess an inherent capacity to form connections to and participate in higher levels of information, at both more expansive and more subtle levels than the finiteness of neurons. Whereas neurons require action (and an absolute 'yes' or 'no' response) from neurotransmitters and electrochemical signaling, isotopes possess properties which may allow for large-scale action without the macros level at which neurons can act.

Isotopes, through their profound role in the structure of the physical Universe, may have a stronger connection to the integer information basis of the Universe (the immutability of the infinite integers and other entities of the Platonic world) than do macros like neurons. As many philosophers have argued, consciousness involves a dichotomy between temporal spatial processes ('here and now') and eternal timeless aspects. The latter refer to such notions as our general statements, logical and mathematical theorems and concepts, etc, which are all *atemporal* entities, meaning that they exist eternally outside of any space–time realm.

To summarize these thoughts, isotopes are physically present as a part of both our brains and the external world. The subtle dynamics of isotopes could have profound effects in neural structures because of the exponentially (and, perhaps, tower-exponentially) high informational potential hidden in isotopic diversity.

## 5.7 Isotopicity in nano and biomedical technology

Most of the ingredients of nanobiotechnology are likely isotopically sensitive, although perhaps to a variable degree. Isotopic effects on some micro-organisms and, to a lesser degree, animals have been studied. For example, there are the effects of deuteration on physiological functions in mice. It appears likely that at such nano interfaces of biology and electronics as biologically inspired molecules and biochips, isotopic purification and purposeful isotopic structuring may result in the enhancement of the performance characteristics of nano-devices.

Another possible line of inquiry regarding isotope effects is medical applications. The fact that physiological response has some degree of isotopic selectivity could potentially be used in 'isotopic pharmacology' to produce isotopically structured or isotopically purified drugs. For example, will mineral vitamin supplements with different isotopic abundances make any difference? For example, natural magnesium is a mixture of three stable isotopes, $^{24}Mg$ (79%), $^{25}Mg$ (10%) and $^{26}Mg$ (11%). If Mg is taken as a mineral supplement, will the action differ if the pill uses only one of these isotopes (purified to a reasonable degree), or mixes them in different proportions? The same could be asked about, for example, calcium, iron, chromium, or selenium supplements, or about other synthetic molecular supplements, of which there are many. Likewise, such quickly developing areas of nano-medicine as nanosurgery, nanobots and drug delivery at the molecular level (e.g. $C_{60}$ 'bucky-balls'—buckminsterfullerene, a ball of 60 carbon atoms—and carbon nano-tubes as drug carriers) are all waiting for research to answer whether isotopicity can provide new options.

As was mentioned above, radioactive isotopes are widely used in medicine (nuclear medicine) for diagnostic and treatment purposes. Likewise, some radioactive isotopes (e.g. $^{14}C$) can be used within the conceptual frame of isotopicity. The latter (by contrast to the standard use of isotopes in nuclear medicine) means some purposeful isotopic placement in specific atomic positions, rather than the bulk use of radioactive isotopes without regard to particular atomic positions. Such placing can (in principle, at least) be achieved using nano-manipulation technologies such as AFM or ion-implantation techniques (Ibrahim and Berezin 1992). Likewise, the electrostatic and radiofrequency traps that are used for laser cooling could also be applied for individual isotopic placement in bio-structures.

Another possible (and likely not yet explored) direction may be 'isotopic nano-bio-mimicry'. Using the fact that some plants and microorganisms show some degree of isotopic selectivity, one can study what physiological and evolutionary advantages such selectivity may bring. Such advantages (if any are identified) could then be imitated in creating purposeful microstructures and micro-devices using the available micromanipulation technologies. Such isotopic nano-bio-mimicry could open up a new avenue for isotopic engineering.

## 5.8 Unstable isotopies and biological information

Here I want to say a few words about non-stable (radioactive) isotopes. Biological activity is usually seen as a chemical level phenomenon, without direct participation from processes inside the atomic nuclei. This may appear to be at odds with common sense, because in reality almost all actual matter resides inside nuclei, not in electronic shells. However, the decays of radioactive isotopes in biostructures, including the human body, can lead to targeted mutations. Most radioactivity in the human body comes from the decays of naturally occurring isotopes of potassium ($^{40}K$) and carbon ($^{14}C$), with half-life times of 1.25 billion years and 5730 years, respectively. Estimates show that in every one of us (using an average human body of 70 kg) there are some 4000 decays of $^{40}K$ per second and about 3000 decays of $^{14}C$ per second (these decay rate numbers for $^{40}K$ and $^{14}C$ are almost the same because while there are far more $^{40}K$ atoms in the human body than $^{14}C$ atoms, the decay time of $^{40}K$ is much longer than that of $^{14}C$). In other words, we humans are constantly being exposed to some natural level of radioactivity coming from our own bodies.

Radioactive decays affect biology in many ways by contributing to mutations ('good' and 'bad') and affecting other physiological and neurological functions. I do not have sufficient knowledge and expertise to discuss these issues in full detail. However, one point that I would like to bring up here is related to my earlier work on positron annihilation (Berezin 1970, 1971). For the radioactive potassium $^{40}K$ (of which there are about 4000 decays per second in a human body) there are three modes of decay, namely beta-decay (electron emission), electronic capture and positron emission decay. Their relative weights (relative probabilities) are 89.28%, 10.72% and 0.001%, respectively. The first two (beta-decay and electron capture) turn $^{40}K$ into a stable isotope $^{40}Ca$, while the (least probable) positron decay converts $^{40}K$ into the (also stable) isotope $^{40}Ar$. Thus, none of these decays change the atomic weight (the number of nucleons) in the decaying nucleus ($^{40}K$, $^{40}Ca$ and $^{40}Ar$ all have 40 nucleons each).

The point that I find noteworthy is that while the positron emission decay is only 0.001% of all $^{40}K$ decays (one out of about 100 000 decays), given that there are 4000 decays of $^{40}K$ per second in a human body, we draw the conclusion that every 25 seconds or so (on average, of course) there is a positron generated somewhere in the body of every one of us. This positron almost instantly annihilates with some nearby electron and the result is the emission of two gamma quanta with an energy of 511 keV (kiloelectronvolt) each (that is the rest energy of an electron—and positron—according to Einstein's relation $E = Mc^2$).

Half-a-million electron volts of gamma quant is quite a high energy. What further secondary reactions and/or processes these gamma quanta can induce in the body

I will not guess at here (perhaps this has been studied?), but it appears that this effect (frequent positron annihilations in the human body, about every 25 seconds) may bring about some non-negligible consequences. Perhaps the processes and reactions triggered by the said gamma quanta can contribute to the flashes of memory and/or intuition that were mentioned above with regard to $^{14}$C isotope disintegrations, a possibility mentioned earlier by Keswani (Keswani 1986).

Another comment to make here is related to the (alleged) randomness of the radioactive decays. Yes, in common sense physics all such decays are supposed to be random. But what does this randomness actually mean and in what sense are these decays random? Can there be some hidden and/or subtle correlations in the positions of decaying atoms and particular instances of their decays? Correlations that are somehow controlled at a macroscopic (body-scale) level? Such non-random triggering of individual decay events could perhaps be biologically controlled by mechanisms similar to those suggested by John Eccles (Eccles 1986).

Within normal decay statistics (fixed average lifetimes), the particularities of individual decays (exact decay moments, relative kinetic energies of electrons and neutrinos and choice of secondary reactions induced by the emitted electrons) can be controlled by a biosystem for its genetic or evolutionary advantages. While for stable isotopes, the informational content amounts to just a few bits per atom (e.g. coded in nuclear spins), for unstable isotopes their quasi-stationary wave functions can support much higher informational content.

If the decaying wave function forms a multiparticle time-dependent superposition of all possible impact-induced events within the mean free path of the emitted electron, the informational content increases even further (perhaps exponentially). Following the logic of the butterfly effect, one could suggest that, in spite of the very low absolute concentration of $^{14}$C, its informational share in biostructures could be comparable to (if not greater than) that of the cumulative total of all the other stable atoms in biosystem. This again shows the power of small variations and small effects (and isotopicity is usually seen as a fringe effect for normal chemistry) to be a source (and sometimes the foundation) of major consequences.

## 5.9 Isotopicity in consciousness dynamics: Anderson localization

*The nitrogen in our DNA, the calcium in our teeth, the iron in our blood, the carbon in our apple pies were made in the interiors of collapsing stars. We are made of starstuff.*

Carl Sagan (1934–1996), *Cosmos*

Among the other topics of this book, I am proposing for consideration some additional ideas regarding the possible relationships of the isotopic diversity of chemical elements with 'soft' areas of science and knowledge, as well as creativity in general. Thus, among the issues that I touch on are such areas as the Platonic world of numbers (IPW), the physics of consciousness, the personal identity problem (the quantum self), holism, homeopathy and even artistic creativity. Almost certainly some readers (again, if there are any readers—something that no author can guarantee) will say, 'what a jumble, how can he put it all together in a single package, and why is he doing so?' To that, following the line of surrealists such as Salvador Dali, I can say that I do what I do and it is up to you (the 'reader'— whether you are real or virtual) to experience any reactions you want (or to experience nothing at all). I, for the most part (especially in later years) see my scientific writings as a kind of 'art' and my take on physics and physical effects has a large component of (what is normally called) metaphysics.

In looking for possible links between isotopicity and consciousness the first thing to note is that isotopic diversity as such, and isotopic fluctuations in particular, can be seen as 'informational drivers' of pattern-forming activity. In that sense, isotopicity serves as an informational source of practically unlimited capacity, somewhat similar to the notion of a 'void' in esoteric mystical teachings. There could be a whole range of ways to suggest plausible quasi-mechanistic models for this 'creative' function of isotopicity. One such model can be argued in relation to the effect of Anderson localization due to isotopic disorder. This model was discussed by the present author (Berezin 1982b, 1984f) in the context of electron localization due to isotopic randomness. Here I restate similar arguments for a more subtle domain of the physics of consciousness.

In favorable cases (e.g. narrow electron bands, strong electron–phonon coupling), isotopic disorder can induce Anderson-type localization of electrons and/or holes. As I suggested in a 1984 paper in the *Journal of Chemical Physics* (Berezin 1984f), isotopically induced Anderson localization may be a possible reason (or, at least, one of the reasons) for the presence of phonon-assisted hopping conductivity in beta-rhombohedral boron, in which the isotopic mass difference between the two stable isotopes of $^{10}B$ and $^{11}B$ is quite significant (10%).

In brief, the model of Anderson localization (Anderson 1958) posits that the disorder in crystals due to the randomness of the location of impurity centers, lattices defects, etc can foster the formation of the trapping centers on which electrons can be trapped in bound quantum states (Berezin 1982a, 1982b, 1984f).

Another ramification of the above theme is the possibility of electronic localization on extreme isotopic fluctuations in a regime when the average-scale isotopic fluctuations leave the system in a sub-threshold state with respect to an incipient localization.

Isotopic diversity leads to a high level of permutational degeneration. In terms of atomic positions, it can be manifested as the quasi-classical degeneracy of having a family of isotopically different spatial arrangements for essentially the same configurational energies. The quantum aspects of this degeneracy may come from the vibrational quantization of a network of chemical bonds, electron–phonon couplings and the coupling of a vibrational manifold with an electromagnetic field (low-frequency).

In this scenario isotopic diversity acts as a symmetry reduction factor that may lead to an additional structuring of the corresponding quantum energy spectrum. No matter how exactly the quantum states may arise from this quasi-classical permutational degeneracy, the degree of the resulting degeneracy of quantum states is so enormous that quantum transitions between states can be executed by virtually any arbitrary weak outside influence. A good analogy to this situation is the 'cat-in-a-laboratory' model, in which the gravitational influence of a cat mixes up the quantum energy levels of a macroscopic system in a laboratory experiment.

In our isotopic version of this scenario various quantum states (in terms of either coherent macrostates or the components of a density matrix of mixed states) could mean various spatial and temporal patterns of localization in a multiparticle system. It is important to note that some models of Anderson localization induced by various sorts of disorder (including isotopic disorder) permit several kinds of nonlinearities, e.g. polarizational nonlinearities, concentrational non-uniformities and fluctuations, etc. This greatly enhances the richness of the possible patterns of charge localization, bond strains, vibrational interatomic activity, etc.

The notion of Anderson localization assisted by isotopic disorder (Berezin 1982b, 1984f) could have a variety of connections to different aspects of mentality. As essentially a symmetry-breaking and pattern-generating mechanism, isotopicity might be related to the self-focusing of mental patterns, transitions between focused and defocused states, oscillatory mental states, etc. Another relevant concept is the model of self-organized criticality suggested by Per Bak. Due to its scale-insensitive character (there is no characteristic length scale) this model might be well positioned to feature the effects of isotopic fluctuations of varying magnitudes and, consequently, various localizational capacities.

## 5.10 Isotopic castling

*Chess is Life.*
　　　　　　　　　Bobby Fischer (1943–2008), eleventh World Chess Champion

From my childhood I liked the game of chess. It is not that I am a strong chess player (most definitely I am not), but chess was (and I hope still is) very popular in Russia and many children learn to play chess quite early in life. And when, some 30 years ago or so, I entered my 'isotopic' period of life, the analogy between isotopes and chess arose from time to time in my mind. Indeed, the diversity of chess figures, all these rooks, bishops, knights, etc resonates (for me, at least) with the diversity of the isotopes of chemical elements.

And when I came up with the idea of isotopic (neutron) tunneling as a model for consciousness (Berezin 1992a), the analogy with 'castling' in chess was almost obvious. The term castling in chess game refers to a special move where the king and the rook exchange positions in a single move. It is only allowed once during the game and some conditions are required for it (e.g. castling is not allowed if the king has already moved or will have to move across a square that is under attack). Now, let us look again at the process of the (alleged) neutron tunneling, keeping the castling analogy in mind.

Let us consider two atoms, two isotopes of the same element, which are different by a single neutron (e.g. $^{12}C$ and $^{13}C$, or $^{28}Si$ and $^{29}Si$). Let us, for simplicity, assume that they are nearest neighbors in a crystal lattice, that is, the distance between them is a few angstroms (say, half a nanometer). These are typical distances between neighboring atoms in crystal lattices. Now suppose that these two atoms interchange (switch) positions in a single correlated jump. Similar correlated tunnel transitions for electrons were analyzed by the present author in 1984 (Berezin 1984a, 1984b, 1984c, 1984d). When two isotopes (two atoms) swap positions, it can be looked upon in two different ways.

(a) It is as if two atoms as wholes do indeed swap positions as two people can swap seats in a movie theater.

(b) 'Atoms' as such do not actually move in their entirety, but rather it is a single neutron that jumps from one nucleus (say, $^{13}$C) to another ($^{12}$C). As a result, the first atom becomes $^{12}$C and the second $^{13}$C. It is something like a money transfer between two bank accounts. But how does this neutron do it, by what 'mechanism'? The most obvious candidate mechanism for such a transfer is quantum tunneling between two potential wells (Berezin (1983), (1984a), (1984b), (1984c), (1984d), Berezin and Jamroz (1984)). In our isotopicity context, the nucleus that can trap an extra neutron makes such a potential well. In this picture we suggest that the neutron can make a direct tunnel jump from one nucleus to the neighboring one. For example, from $^{13}$C the neutron jumps to a neighboring $^{12}$C nucleus, so that the first atom becomes $^{12}$C and the second one becomes $^{13}$C. Thus, the two atoms exchange their isotopic identity without moving as wholes.

However, the problem with such a picture is that a straightforward quantum calculation of such a probability (and for that we need to know the binding energy at both wells and the tunneling distance) results in an extremely low probability of such an event (tunneling) occurring spontaneously. Hence, in order to support the plausibility of such a process (process (b)), we need to offer some superseding mechanism that greatly (exponentially) enhances the tunneling probability over and above what a direct tunneling formula gives us for the tunneling rate. Perhaps this might be an Eccles effect (Popper and Eccles 1977, Eccles 1986), or perhaps the key to this enhancement mechanism can be sought at sub-nuclear scales (maybe at scales of the Planck length?), or then again perhaps it is some kind of an information-driven effect where the particle (neutron) exercises some 'personal interest' in getting through the barrier (Berezin and Nakhmanson 1990)—no matter how fantastic such suggestions may look from the side of the mainstream physics. Perhaps there might be some other hypotheses (or, if you wish, speculations) along this line, but at this point I am leaving this topic as an open-ended outstanding challenge.

I would like to say a few more words to explain more clearly what I mean by the above-mentioned Eccles effect. Here is a (rather extended) quotation phrased in the terms of the idea of 'contingency' from the book *Mysticism and the New Physics* by Michael Talbot.

[It was] Willard Gibbs, who first proposed that the universe was contingent (predictable only within statistical limits) as opposed to deterministic. As early as the 1870s Gibbs was formulating his ideas of contingency. There is an overwhelming probability that every time you hit a billiard ball from the 'same' direction with the 'same' amount of force it will react in the 'same' way. But there are fringe occurrences—oddities on the frayed borders of our cause–effect reality—that indicate this contingency of the universe. In a contingent universe, although the billiard ball will react in the 'same' way most of the time, there is a chance that it will not react, or even do something totally unpredictable. According to Gibbs's idea of contingency the physicist can no longer deal with what always happens, but only with what happens an overwhelming majority of times.

(Talbot (1992), p 19)

The idea that the probability of quantum mechanical tunneling may be (exponentially) enhanced by some 'outside' interaction—this time we are talking about consciousness (Berezin 1992a)—is precisely what Talbot refers to as 'fringe occurrences—oddities on the frayed borders of our cause–effect reality'. No matter how strange such an idea may look, the Eccles effect in this case (Berezin 1992a) means that the coherent conscious activity 'instructs' a particular neutron to 'disregard' the standard (exponentially small) quantum tunneling probability and to perform such a (quantum-almost-improbable) jump from, say, one ($^{13}$C) nucleus to another ($^{12}$C) nucleus. In other words, consciousness 'allows' particles at the atomic level to circumvent (ignore) quantum probability limitations and to commit tunneling acts much faster (exponentially faster) than would otherwise be the case.

## 5.11 Neutron tunneling and quantum consciousness

*We have a closed circle of consistency here: the laws of physics produce complex systems, and these complex systems lead to consciousness, which then produces mathematics, which can then encode in a succinct and inspiring way the very underlying laws of physics that gave rise to it.*

Roger Penrose (b 1931)

A common feature of a number of versions of the quantum theory of consciousness is that they assume a two-way linkage between consciousness and the physical world. In other words, consciousness as a quantum phenomenon and the surrounding physical reality are jointly locked into an interactive loop. Some recent experiments seem to confirm the direct action of consciousness on physical systems, such as electronic random number generators (e.g. Jahn and Dunne 1988). A tentative conclusion from these experiments is that there is some kind of 'cooperation' (resonance) between the physical system and the consciousness of an observer.

Extending these ideas to the case of the homeopathic protocol (section 5.16), one could suggest that the conscious intention of the observer (in this case a doctor) forms an interactive informational loop with the homeopathic remedy under preparation, provided that it is prepared within the restrictions of a properly applied protocol. In some sense, a patient could also be included in this informational interaction.

The next step in our model is related to the memory-forming mechanism based on dynamical isotopic patterning (Berezin 1992a). As we mentioned earlier, the density of isotopic information storage is, in principle, of the order of atomic density (or one of two orders of magnitude below this, to assure robustness and redundancy). This amounts to about $10^{20}$ bits $cm^{-3}$. At first glance, this kind of information storage appears to be positionally inflexible and non-dynamic. In order to argue otherwise, let us consider two different isotopes occupying neighboring sites in a solid or molecular structure. Take both sites to be energetically equivalent. A permutational exchange of two isotopes between both of these sites is possible as a quantum mechanical tunneling process (quantum ring diffusion, or 'isotopic castling', see section 5.10). The rate of such processes, $1/T$ ($T$: characteristic time), can be written in a way that is similar to the case of a radioactive decay. In common symbols, it is a product of an exponentially small tunneling factor, $\exp(-W)$, and an attempt frequency, $f$, according to the equation $[1/T = f \cdot \exp(-W)]$, where $W$ is a proper combination of geometrical and energetic parameters of a given quantum system (more details on this can be found in any introductory text on quantum mechanics).

The meaning commonly ascribed to the above equation is statistical: at every attempt the system has an equal probability, proportional to the tunneling factor $\exp(-W)$, to exercise a transition. In the case of a tunnel exchange of sites by isotopes, the attempt frequency is defined by the Heisenberg delocalization rate. For a typical atomic mass $M$ (about, say, $1/10^{25}$ kg) this rate is of an order of $(x^2) \cdot M/h$, where $x$ is a characteristic confinement length and $h$ is the Planck constant. A typical inter-atomic distance ($x$) is about three angstroms. This leads to an attempt rate that is comparable to typical phonon frequencies in condensed matter physics (about $1/10^{12}$ s).

The above values mean that isotopic rearrangements could proceed at very fast timescales (e.g. picoseconds), provided that a certain physical condition can be imposed on a system. This condition enhances the tunneling factor (even for a short time interval) from 'normal' (exponentially small) values to values comparable to unity. Such a condition can be seen as an analogue to the establishment of coherent resonance tunneling in the entire system (overall quantum transparency). We postulate that it is a state of conscious activity that imposes such a condition and as a result releases the occurrence of multiple fast isotopic permutations in apparent violation of the standard quantum tunneling rates.

In ordinary (non-coherent) conditions such enhancement normally does not happen. The reason for this is that these exceedingly small tunneling factors drastically suppress the observable isotopic exchange rates at low temperatures despite a quite high Heisenberg delocalization rate $(x^2) \cdot M/h$. Assuming, alongside Roger Penrose's ideas (Penrose 1994, 1996), that gravity-induced quantum reductions can be informationally selective ('catalyzed' by specific information), we arrive at a dynamic quantum model of consciousness in which such goal-oriented (in terms of arriving at a 'meaningful' atomic pattern) isotopic orderings are interactively linked to gravitational quantum reductions.

Adding to the above, here I want to outline a few other mini-hypotheses regarding possible relationships between isotopicity and consciousness for possible exploration by some present or future readers. In brief, they are:

(1) the existence (and/or formation) of consciousness-driven positional correlations among $^{13}$C atoms;

(2) the motion (hopping) of $^{13}$C via enhanced neutron tunneling, where such a process could perhaps be seen in analogy to the so-called 'anti Zeno-effect' (the quantum amplification of tunneling transitions);

(3) the possible optimization of the concentration of quantum computing-active $^{13}$C atoms above their standard isotopic abundance;

(4) the diversity and interaction of the characteristic timescales for the operation of $^{13}$C-based quantum computing (perhaps a broad range of scales);

(5) the influence of the quantum computing dynamics of $^{13}$C on consciousness;

(6) the possibility that $^{13}$C-based quantum computer systems can (or might) operate 'above' the level of 'regular' consciousness (perhaps a Jungian sub-/super-consciousness);

(7) isotopicity as a connector to the universal library of patterns (the Platonic world); and

(8) the self-stabilization of coherence in $^{13}$C (sub)systems.

Undoubtedly, some other possible links can be identified and indicated. Here I have provided only the most basic and it is inevitably a vague list. Some of these questions are, in principle, experimentally addressable through the shifting of isotopic abundances and/or other facets of isotopic engineering.

## 5.12 Isotopicity and the Gaia concept

*How inappropriate to call this planet Earth when it is clearly Ocean.*
Arthur C Clarke, 1917–2008 *Nature* (8 March 1990)

Now I want to move my discussion of isotopicity and quantum physics towards grosser (cosmic) scales. In 1999 I sent a short letter (Berezin 1999) to *Discover* as a comment on an article about the Gaia hypothesis. The Gaia hypothesis is generally associated with the name of James Lovelock, although the origins of it can be traced back to earlier thinkers, for example, Vladimir Vernadsky (1863–1945) and Pierre Teilhard de Chardin (1881–1955).

This is what I wrote in my letter to *Discover*:

It is exhilarating that, despite scorn from the mainstream science establishment, the Gaia hypothesis is alive and well ('Is the Earth alive?', October). As noted, 'powerful metaphor never relies on only one meaning' hence it is befitting to extend the Gaia concept to another self-regulating system that is central to our existence—the sun. Is the sun alive in a Gaian sense? Probably.

Worship of a 'living' sun was persistent throughout history. Egyptian Pharaoh Akhenaton (1353–1335 BC), who introduced the sun as sole deity, is but one example. Such views mesh well with the 'modern' fact that the prime chemical elements of Earth's biology (hydrogen, oxygen, carbon, and nitrogen) are the same elements whose nuclear transformations are responsible for the sun's energy production. Is this coincidence or evidence of some fundamental biological communality?

As discussed in many papers and books (e.g. Berezin and Nakhmanson 1990) the idea that 'most things are alive' is strongly embedded in many cultural traditions. Recently, Gregory Sams (Sams 2009) moved the idea of a living Sun to the level of physical hypotheses, arguing for it on the basis of the theories of chaos and self-organization. In his view all stars are 'alive' and form communities of living beings—what we call galaxies. In his words, galaxies do not fall apart because 'they are smart' and are held together by internal bonds of cooperation (like cells in the body), rather than by such (hypothetical at this point) things as so-called dark matter. Isotopicity has a role to play in this context, if for no other reason, than for the fact that energy production in stars is based on thermonuclear reactions between isotopes.

For those who may decide to dwell further on the issues of isotopicity, biology and evolution, here is a list of possible themes. They are all in the form of open-ended questions ('quests'). Some of them originated during conversations with friends and I in no way consider this list to be exhaustive or complete, on the contrary, the readers of this book may well add much to it. Some of these questions may have some import for biomedical thinking and may lead to possible experimental studies.

(1) Isotopic randomness as a factor (driver?) of evolution, including both overall (ascending) evolution and evolution within species.

(2) Will a minor departure (say, by 1 or 2%) from standard isotopic abundance (say, between $^{14}$N and $^{15}$N) lead to health problems? If so, can this be quantitatively graded?

(3) The propagation of solitonic waves (localized vibrations) in biological structures. Because the propagation of vibrations is mass-sensitive, isotopic effects here may be quite significant, up to, perhaps, the formation of singularities (in two-dimensional systems pulse propagation is a singularity-type effect).

(4) Isotopic differences in bosonic and fermionic statistics. Different isotopes of the same element may have different quantum statistics. For example, $^{12}$C is a boson, while $^{13}$C is a fermion. Can we talk about bosonic and fermionic DNA (or segments of DNA) that may be formed by fluctuations in isotopic distribution? (Such fluctuations may arise spontaneously or may be set up by outside factors like deliberate isotopic manipulation.)

(5) Aging (at the cellular and organism levels)—do we have evidence of a shifting of isotopic abundances? If so, can such a fact be used to inhibit or reverse the aging process?

(6) Do decays of trace radioactive isotopes (primarily $^{14}$C and $^{40}$K) affect (accelerate?) the aging process? Or, perhaps, a contrasting argument can be

raised: could an *increase* of such isotopes turn out to act as a beneficial factor, similar to the action of the supposed hormesis effect (the strengthening of the immune system by radiation)?

(7) In water the nearest-neighbor ordering (the range of the positional correlations) is about 50 angstroms. This is about the same scale as the isotopic fluctuations in DNA. Can we draw some consequences from near- and long-range isotopic order in DNA and other biologically important molecules?

## 5.13 Isotopicity and personal identity

*All theory is against the freedom of the will; all experience is for it.*
Samuel Johnson (1709–1784)

Within the paradigm of isotopicity one can pose a number of questions that are pertinent to the role of isotopic diversity in various aspects of biology. We can refer to the quantum foundations of biological evolution, the emergence of consciousness and the (so-called) 'mind/body problem', with its multiple ramifications for the 'physics' of spirituality, creativity, health states and healing techniques, human interconnectedness and related issues.

Now, a few words on possible links between the informational aspects of isotopicity and biology and the notion of the (so-called) 'quantum self' (Zohar 1990). Namely, we can look at isotopicity as a possible contributing factor to the quantum foundations of human individuality and self-awareness and the nature of personal identity (Berezin 1994a).

Many of us often wonder about the enormous complexity and diversity of our world. However, all the stars and galaxies, crystals and minerals, and living creatures of all sizes and types, including ourselves, are made from 'just' several dozen chemical elements (and their isotopes, of course). Nature appears to us as an unbelievably creative, self-propelled and self-motivated enterprise. Furthermore, this enterprise seems to be working on a very limited resource base (just atoms and physical fields and not much else). And yet, it never falls into pessimism or recession. It tirelessly devises all its new structures and living beings at an enormous pace, but also with (seemingly) tremendous wastefulness.

Furthermore, there seem to be no truly impenetrable barriers between 'natural' creativity and the creativity of human beings, whether the latter is at an individual level or at the level of collective social activity. There are several major (and many more less common) ways and traditions for people, more or less systematically, to express their connectedness to the whole Universe and to proclaim their own place and role in it (Abraham 1994). Such ideas as that of a supreme god (all the major monotheistic religions), god-in-nature (Albert Einstein), or various polytheistic gods and goddesses, and concepts like Gaia (James Lovelock) or the participatory Universe (John Archibald Wheeler), etc, are all the examples of our indestructible desire to be inherently related to the 'whole' and the 'eternal'. Even within the most extreme materialistic doctrines of a purely physical random Universe ('only matter and no spirit'), one can usually also observe some remnants of this search for universalities. This can take the form of, say, appeals to such concepts as universal physical symmetries and primordial quantum vacuum, or the models of an intrinsically interwoven network of baby Universes in recent cosmological theories, and many other similar constructions.

Yet regardless of which of these different (and often opposing) metaphors each one of us uses to express our embedding in the Universe, there seems to be a common inference for all these views. This common ground can be formulated as the notion that we somehow belong to the Universe and are produced by it (either for a purpose or at random—if we want to use these terms). Consequently, everything that is created by us, as individuals or as an entire human civilization, can be claimed by the Universe as belonging to the interconnected totality of its creative dynamics.

Let us now look at the specific tools available to Nature at the level of Earth's biology. Life on Earth, at least as we understand it now, is based on chemical diversity. All the objects that surround us, including ourselves, are made up of just a few dozen (about 80) different chemical elements. The enormous variety of their possible combinations accounts for all the richness of minerals and other non-organic structures found in nature. Even more impressive is the almost unlimited variety of biological structures and biochemical processes that are responsible for the existence and evolution of all living beings on this planet.

Out of all the chemical elements only four are absolutely critical for all of Earth's (known) biology, and these are hydrogen, carbon, oxygen and nitrogen. Let us call this group of elements the 'HCON' group. What is interesting (and perhaps, peculiar) is that the HCON elements are also thought to be responsible for the energy production cycle in the Sun. In this way we humans and the Sun may be said to be 'relatives' (Sams 2009).

In 1938 Hans Bethe (1906–2005) and Carl von Weizsacker (1912–2007) suggested (independently) that the so-called CNO-cycle is a catalytic mechanism through which the Sun fuses hydrogen into helium, and that this is the source of the energy generated by the Sun. This reaction goes through the chain (a loop) of isotopes of oxygen, carbon, nitrogen, as well as hydrogen (primary 'food') and helium (end 'product'). The heavier nucleus (C, N, O) serves as a reaction site for the hydrogen atoms (protons) to turn into helium. This isotopic reaction looks like this:

$$^{12}C \rightarrow {}^{13}N \rightarrow {}^{13}C \rightarrow {}^{14}N \rightarrow {}^{15}O \rightarrow {}^{15}N \rightarrow ({}^{12}C)...$$

The by-products of this loop of isotopic nuclear reactions are positrons, gamma rays and neutrinos, and the total energy yield of one loop cycle is 26.8 MeV. After six stages of nuclear reactions we arrive back at the original $^{12}C$ nucleus and the loop repeats over again. In this scenario the carbon, nitrogen and oxygen isotopes are actually one nucleus that goes through a number of transformations in a repetitive catalytic loop.

It should be noted that at the time of writing (2016), it is thought that the CNO cycle is only responsible for about 1.7% of the energy output of the Sun, with the rest coming from the proton–proton fusion reaction. Nonetheless, despite the fact that the CNO cycle only makes up about 1/60 of the total energy output (according to current science—this number well may be revised in future), we still can see ourselves as 'isotopic relatives' of the Sun.

Apart from these major elements (H, C, O and N), a number of other elements are also essential for our biology (e.g. phosphorus, sulfur and sodium), together with the (even larger) group of 30 or so elements loosely labeled 'microelements'. Elements from the last group (microelements) perform some more specific functions, often with different degrees of significance for various life forms (e.g. elements like magnesium, copper, iron, calcium, zinc, etc). The biological role of a few other remaining elements (e.g. rare earth metals) still remains somewhat unclear. However, realizing how creative and inventive biological Nature is in terms of utilizing all available resources, it appears quite likely that it finds ways to make use of all (or almost all) the elements of the periodic table for some specific or focused biological functions. And—extending such an argument—if Nature utilizes the diversity of chemical elements, does it also employ the targeted use of isotopic diversity for its creative endeavors?

In other words, it turns out that almost all the chemical elements play some role in living organisms. Even such rare and exotic elements as gold and uranium are known to accumulate in some specific organs and tissues in concentrations that significantly exceed their average chemical abundance in the Earth's crust. Therefore, one can conclude (at least as a statement of plausibility) that living Nature is 'smart enough' to find some use for (almost) all the members of the periodic table of elements. But if this is so, a similar question can be asked about the isotopic diversity of chemical elements.

About 3/4 of the chemical elements have at least two (and often three or more) stable isotopes. For example, 99% of all natural carbon atoms are $^{12}C$ atoms. The nucleus of the $^{12}C$ atom consists of six protons and six neutrons. However, this 1% still amounts to a very large absolute number of C atoms that are $^{13}C$ isotopes. The latter has six protons and seven neutrons in its nucleus and, consequently, is about 8% heavier in mass than the ordinary ($^{12}C$) carbon atom.

Likewise, oxygen has three stable isotopes: $^{16}O$ (99.8%), $^{17}O$ (0.04%) and $^{18}O$ (0.2%). Again, despite a seemingly small fraction of minority isotopes of oxygen, even such a 'negligible' concentration as 0.04% translates into a very impressive absolute concentration of $^{17}O$ atoms. One of every 2500 oxygen atoms is an $^{17}O$ isotope. A tiny living cell of, say, 1 μm size has about $10^{10}$ (ten billion) to $10^{11}$ (a hundred billion) oxygen atoms (mostly as water molecules). Therefore, such a cell still contains several million minority $^{17}O$ atoms. Being placed among oxygen atoms, these minority atoms can form a tremendous number of spatial combinations. The latter can make a difference in a number of ways. For example, $^{17}O$ atoms have non-zero nuclear spin and, therefore, their magnetic properties differ from those of $^{16}O$ and $^{18}O$ isotopes, with the latter having zero nuclear spin. Therefore, the response to magnetic fields of various origins and spatial scales can be isotopically sensitive. At the micro level such micro-magnetic fields can originate from some micro-electrical currents, which, in turn, may be due to ion transport at a cellular level.

The difference between majority and minority isotopes is not always clear-cut. Some chemical elements have two (or more) isotopes with comparable abundances. For them there may not be a clearly designated majority isotope. An example of this is silver, which has two stable isotopes with almost equal abundances, namely $^{107}$Ag (51.8%) and $^{109}$Ag (48.2%). A 'champion' of polyisotopicity, tin (Sn), has ten stable isotopes.

So, once again, anyone can repeat the above stated question in the following form: if Nature is 'smart' enough to make such impressive and skillful use of chemical diversity, why would it make no clearly identifiable use of isotopic diversity? To repeat from section 3.9, physicist Murray Gell-Mann (Nobel Prize in physics, 1969) has a quotation attributed to him that says 'Anything which is not prohibited is compulsory' (Comorosan 1974). This means that everything that is possible does indeed happen, somewhere, some time; even the Narnia of C S Lewis or jinn coming out of bottles. And as for flying carpets and witches flying on brooms, perhaps there may be a way to shield against the effects of gravity, or to use some (not yet discovered?) effects of anti-gravity for that matter. So, in the context of this book, Gell-Mann's principle pretty much supports the idea that Nature may indeed be looking (hopefully successfully) to utilize isotopic diversity for a variety of ends.

Stable isotopes of the same chemical element differ in their masses because they have different numbers of neutrons in their nuclei. This mass difference leads to detectable variations in the rates of some chemical reactions, atomic diffusivity, etc. But mass difference is not the only route through which isotopic diversity affects the kinematics and dynamics of physical and chemical processes. Isotopes also differ in their nuclear magnetic moments (nuclear spins) and (due to a combination of mass, nuclear spin and nuclear size variations) have slightly different positions for corresponding atomic energy levels (eigenvalues). Such differences (isotopic shifts), although small, may nevertheless be important in various resonance-type phenomena which can 'amplify' these small differences. Isotopic shifts have been well studied in atomic and molecular spectroscopy. Isotopic variations of physical and chemical properties are successfully exploited in several existing isotope separation technologies.

## 5.14 Isotopes: a 'secret tool' of Nature's creativity?

So, in view of all the above, one can ask a devil's advocate question: why, up till now, has there not been much evidence that Nature does indeed use isotopic diversity for some significant biological ends?

My, albeit tentative, answer to such a query includes several arguments.

Nature *does* in fact use isotopic diversity and indeed uses it in quite extensively 'all the way through', but we have been overlooking this for several reasons. The two major reasons for the biological significance of isotopic effects being 'veiled' from the mainstream of contemporary bioresearch could be as follows.

(A) Isotopic effects are *subtle*. In other words, they are masked and obfuscated by the much stronger, more pronounced and more obvious effects arising from chemical diversity.

(B) The impact of isotopic effects on biological dynamics is primarily of an informational (and not an energetic) character.

The first point to mention here is the issue of the subtlety of isotopic effects. This is essentially the issue of the existence of two grossly different levels of intensity for the underlying effects: chemical effects versus isotopic effects. Chemical differences work primarily at the level of atomic bonds. Chemical bonding is governed by strict rules of valence considerations and elemental compatibility in forming compounds. Everyone even marginally familiar with chemistry knows that not all combinations of elements form compounds, compounds rather form in a way that strictly follows the numerology of atomic combinations (e.g. it takes strictly two hydrogen atoms, one sulfur atom and four oxygen atoms to form one molecule of sulfuric acid, etc). There are no similar restrictions for isotopic ratios; isotopes of any given element can be mixed in any arbitrary proportion without upsetting chemical bonding regulations.

Furthermore, crystalline structures have very rigid arrangements in terms of chemical bonding and allowed combinations of different chemical elements. For instance, crystalline quartz, $SiO_2$, has a well-defined structure that contains silicon and oxygen atoms at specific lattice sites. The positioning of these lattice sites is quite inflexible and is predetermined by chemical 'rules and regulations'. That is, if an atom of Si or O is in a 'wrong' location, it is seen as a lattice defect, as an imperfection of the lattice structure. In 'good' crystals the concentration of such defects is small and, in principle, in 'ideal' crystals they (defects) can be eliminated altogether (one could say that it is theoretically possible to grow a chemically 'perfect' crystal).

This is not so for isotopes: even within a fixed chemical structure, the positioning of various isotopes of the same chemical element remains a matter of pure chance. There is an isotopic randomness within the chemically predetermined structure. We can refer to this as the concept of isotopic freedom (section 4.5). Isotopes can be freely moved and rearranged within chemically fixed structures. This leads to microscopically distinguishable patterns that are not subjected to statistical restrictions because of the Gibbs paradox (Jaynes 1992, Allahverdyan and Niewenhuizen 2006). Furthermore, such isotopic patterns can be informationally loaded. This means that such a pattern can store some externally input information which may be coded in a specificity of isotopic distribution (Berezin (1984h), (1992b), Pui and Berezin (2001)). This applies not only to solid crystal lattices, but also, to some degree, to quasi-crystalline structures of liquid systems (Berezin 1986b, 1990a, 1990b, 1994c).

In addition, isotopic information storage may have some relevance for issues of alternative biology and evolution. One could suggest that isotopic diversification at the microscopic level bears some 'panpsychistic' connotations (Berezin (1984g), (1987c), Berezin and Nakhmanson (1990)). At the level of brain function, microscopic isotopic patterning may be relevant to physically fundamental aspects of the personal identity problem (Berezin 1994a).

Our last remark is related to the specific quantum dynamics that isotopicity may 'choose' as a route to contribute to the emergence of the quantum self (Zohar 1990, Berezin 1994a). This applies both to our (human) level and the level of unanimated (e.g. crystalline) objects. This is the new science of quantum computing (section 4.13).

Quantum computing, or the theory of quantum automata, is a new segment in theoretical physics. Its emerging importance is quite interdisciplinary and may far transcend the physics of computational systems in the proper sense. It relates such apparently quite distant things as cosmology, the nature of time, evolution, the emergence of information in non-equilibrium non-linear systems and the physics of consciousness.

Generally, a physical computation is a process which produces outputs that depend in some desired way on given inputs. The nature of inputs and outputs should normally have some informational interpretation. This generalized vision admits under the notion of computing such processes as, e.g. ontogenesis, bio-evolution, the operation of the immune system and many other bio-related processes. The latter may include consciousness or some steps of the homeopathic protocol involving mind–matter interactions (Berezin 1990b, 1994c).

Isotopicity, as a level of informational diversity that is additional to common chemistry, could be a missing link between the requirements of conscious processing and a quantum measuring system operated on the principle of quantum super-positions of alternative states (outcomes). The combination of isotopicity and quantum computing proposes viewing isotopically patterned molecular or crystalline structures as a kind of quantum mechanical measuring system or quantum computer. The fundamental capability of isotopicity to perform such a task lies in its information potential, as was explained above using a 'freedom-within-fixed-structures' paradigm.

While it seems unlikely that there could be a single all-purpose definition of 'creativity', we can, nonetheless, offer some functional definition befitting the context of this book. In the human realm creativity is generally understood as the generation of new patterns from existing 'building materials'. All books (and as a corollary—all ideas) that can be written down use the letters of the various languages' respective alphabets, which (with the exception of Chinese) constitute quite small sets of characters (some 30 to 50 in most cases). Furthermore, since everything can be digitized, any information (pattern, picture, etc) can be converted into a binary string consisting of only two characters (e.g. 0 and 1). As was mentioned above, any information, no matter how rich and complicated, can be coded by a sufficiently long integer number. For this, so-called Gödel numbering is especially handy. This is the type of numbering (Gödel numbers) that Kurt Gödel used in his famous undecidability theorem.

As for the physiological causes of human creativity (which, of course, are very complicated and multiple), we can invoke here some possible (hypothetical) isotopicity component in them. As was mentioned above (sections 5.8 and 5.13), random (or, perhaps, correlated?) radioactive decays in the human body (including the brain) could, perhaps, trigger flashes of intuition, creativity, extra-sensory perceptions and other similar effects.

Many recent developments indicate that we are presently well along the path to eventually creating super-intelligence. Of course, the chess programs that currently exist, and which have already beaten even the best human chess masters, cannot yet be classified as super-intelligence. However, the very existence (and advancement) of such programs is a strong indication that further developments may soon lead to the inclusion of other human mental capacities in computer programs. More and more human features (such as object recognition and guessing) can be imitated, often with superior performance to actual humans. Were do we stop or should we be stopping at all?

Another pertinent claim is the so-called 'Flynn effect', according to which the average human intelligence quotient (IQ) increases gradually over the years. While the whole notion of IQ is not without controversy, the said effect (if real) can be attributed to several factors. Among these are better nutrition, better education and advancements in global information technology. Thus, if the growth of IQ scores can be taken at face value, it could, in principle, generate a 'blissful circle', in which more and more 'smart' humans can build smarter and smarter information systems, which, in turn, can foster the further growth of intelligence in humans.

Indications of the potentially self-enhancing human–technology loop can be traced back to the earlier literature. As early as 1793, the French philosopher and mathematician Antoine-Nicolas Marquis De Condorcet (1743–94) was discussing the growth of human capacities as a result of the progress of science: 'At the present time a young man on leaving school may know more of the principles of mathematics than Newton ever learnt in years of study or discovered by dint of genius, and he may use the calculus with a facility then unknown' (Condorcet 1955, p 196).

And while these words where written over 200 years ago, our digital age amply confirms them. However, it should be kept in mind that the increase of 'measured' (by whatever means; IQ scores or any other tool) 'intelligence' should not be confused with the growth of some far more important human faculties, such as comprehension, compassion, ethical values and responsibility. And on the latter side, the developments (if any), in my view at least, remain minimal and more often than not they are in fact negative. Very few people are looking forward with optimism to the coming 'robot age'.

## 5.15 Is Nature 'isotopically smart'?

*Simply because you do not have evidence that something exists does not mean that you have evidence that it doesn't exist.*
> Donald Rumsfeld (b 1932), former US Secretary of Defense

For the curious reader who takes the time to go through the arguments presented in this book, the prime puzzle may likely be spelled out this way: if isotopicity (meaning the effects of the diversity and randomness of stable isotopes) is as universal and omni-penetrative as this author suggests, why then has nobody (or, let me say more carefully—*almost* nobody) raised this issue to the point of a well-formulated 'quest', either theoretical, or experimental, or both? Where does the crux of this omission lie— in the research community, or, maybe, in Nature itself? Or, perhaps, a bit of both? Who (or what) is not 'smart' enough here?

Let me first of all advocate on the side of Nature and primarily on the issue of isotopic biology—the concepts presented above in this book. To address this, and to repeat what has been said several times above, I propose that the prime philosophical question behind the concept of isotopic biology can be formulated in the following way. If Nature is smart enough to use the diversity of chemical elements for biology (almost all the elements from the periodic table have some biological functions, including microelements), then it may look somewhat odd that Nature would omit to use such a mighty additional informationally rich resource as the diversity of stable isotopes for the structuring and functions of biological systems at all levels of evolution and complexity. The likely 'answer' to such a 'puzzle' is that, yes, Nature most likely uses it (isotopic diversity) but we so far failed to detect it and even (largely) failed to look at this even at the level of hypothesis, not to mention any targeted experimentation. One of the prime aims of this book is to draw the attention of the world's research community to this incipient research area of stable isotopicity and isotopic engineering—a direction that (with some luck) may turn out to be a newly found gold mine for physics, biology, biomedicine, material science, cognitive sciences and informational technology in a broader sense.

Perhaps one way to test Nature's smartness with regard to isotopic diversity in biology is to set up some experiments on isotopic replacements in bio-systems. In fact, some such experiments were reported and here I suggest a few more possibilities.

For example, experiments on isotopic replacements for elements heavier than hydrogen have been reported for $^{13}$C and $^{18}$O. Both experiments succeeded in replacing about 60% of all body carbon or oxygen in mice with $^{13}$C or $^{18}$O (their normal isotopic abundance is 1.1% and 0.2%, respectively). No obvious pathological changes were found in any of the organs. However (at least, to my best knowledge— and I do not guarantee its completeness), there have been no specific studies on how such isotopic metamorphosis affects various physiological and/or psychologically conditioned responses. Several suggestions for such studies are listed below. Although they are written in the context of physiological rather than psychological experimentation, some or all of them could be extended to psychology-related studies. These suggestions were proposed by the present author over 25 years ago (Berezin 1990b) and I have not checked exhaustively in the more recent literature to see if any of them (or others) have been experimentally tested.

1. *Combinations of isotopes.* An obvious extension of studies on $^{13}$C and $^{18}$O mice (or for that matter simpler organisms such as *Drosophila*) would be a combination of two or more stable isotopes in variable ratios. For instance, the cumulative (synergetic) effect of a triple replacement of $^{12}$C by $^{13}$C, $^{16}$O by $^{18}$O and $^{14}$N by $^{15}$N could lead to much stronger effects than any of them separately.

2. *Thermal regulation.* Isotopic replacements in C, O and N from their standard isotopic abundances may change the densities of all structures by several per cent. Diffusivities and other kinetic characteristics, especially those related to quantum tunneling, could change even more strongly. Thus, isotopic replacements might lead to the 'fine tuning' of thermal regulation mechanisms in mammals. Similarly, isotopic shifts might change the speed of the propagation of nerve impulses and other related characteristics.

3. *Magnetic effects.* Magnetic effects in biology are generally considered to be subtle, but they are of importance in some specific cases (e.g. in the orientation mechanism of birds and the responses of some microorganisms to magnetic fields, etc). In view of the differences between $^{12}$C and $^{13}$C's nuclear magnetic moments (see above), magnetic experiments on $^{13}$C organisms might lead to non-trivial insights.

4. *Isotopic fluctuations.* Any carbon chain contains about 1% of $^{13}$C atoms. This, in fact, is a very large concentration in absolute terms. If a similar concentration related to an ordinary chemical impurity in a crystal, this could change its properties in a drastic way. For isotopically blended crystals such properties as the Rayleigh scattering of light on isotopic fluctuations, or modification of the thermal or electrical conductivity, might show detectable variations (Berezin (1988b), Berezin and Ibrahim (1988), (1991)). The same applies to the minority isotopes of oxygen or nitrogen in living structures. In living matter the performance of DNA might be affected by the positional fluctuations of minority isotopes. This, in turn, could contribute to mutability and, consequently, one can suggest that isotopic randomness could act as an additional genetic factor. Fluctuations are generally proportional to the square root of the number of particles. Therefore, the effective (average) size of isotopic fluctuations will be appreciably different if, say, the concentration of $^{13}$C is changed from 1% to 0.1%. (In this example, the spatial size of the fluctuation will increase $10^{(1/3)} = 2.15$ times if the distribution is uniform in three dimensions; for low-dimensional systems, such as carbon chains, the size scales differently.) One might think, therefore, of some kind of 'inverse' experiments to those done with $^{13}$C and $^{18}$O mice. The use of major isotopes ($^{12}$C, $^{16}$O, $^{14}$N) purified from the admixture of minor isotopes could perhaps suppress fluctuations and result in 'improved' genetic stability ('isotopic eugenics'?). Similarly, this might lead to some variations in neurological processing, memorization and other perception- and consciousness-related activities.

5. *Positional correlations.* As a supplement to 'isotopic indeterminism' (Berezin 1987c), it is also conceivable to think of possible positional correlations of $^{12}$C and $^{13}$C atoms during DNA replication. Namely, does a secondary spiral have any trace of the individual $^{12}$C–$^{13}$C pattern of the primary DNA spiral? The establishment of even a weak positional correlation would certainly be of interest, and experiments with organisms enriched by $^{13}$C to a variable degree (i.e. 10%, 20%, etc) could provide a key for such studies.

## 5.16 Isotopic ordering in liquids and 'soft structures'

*Science fiction is any idea that occurs in the head and doesn't exist yet, but soon will, and will change everything for everybody, and nothing will ever be the same again. As soon as you have an idea that changes some small part of the world you are writing science fiction. It is always the art of the possible, never the impossible.*

Ray Bradbury (1920–2012)

The following story may be the strangest part of this book. Those (hypothetical) readers who may proceed to read the book will most likely find that the matter I am talking about below is utterly baffling and unorthodox. Or, perhaps, it lies at the borderline of science fiction, as the above quotation by Ray Bradbury may imply. Yes, it is nothing less than a scientific attempt to propose a physical model for homeopathy, one of the most controversial and hotly disputed practices of (so-called) 'alternative medicine'. In fact, to avoid upfront accusations of excessive claims, I must right away make it clear that my purpose here is far less ambitious.

In brief, my work in this direction was to propose the idea that the (claimed) memory effects in water might be related to the isotopic self-structuring of (oxygen) isotopes in a liquid matrix of water. This idea (and related to it the proposition that 'healing crystals' form information-carrying 'isotopic neural networks') was developed by me over a number of years (from 1987 to 1994) and was outlined in a number of publications, which are referred to in the references of this book.

No, I will not go on to discuss the whole 'art and science' of homeopathy or anything on the practical side of it. There is no shortage of literature dealing with the practical aspects and possible theoretical foundations of homeopathy (e.g. Bellavite and Signorini 2002). I am no way a medical doctor or even a 'medicine man', as this term is traditionally understood. What I actually proposed in some papers published in the early 1990s was the isotopic model for the memory effect in water. If *that* can be substantiated (or at least plausibly argued) on the basis of isotopicity (isotopic diversity), then others, who are specifically interested and better versed in the actual art of homeopathy, can use these ideas as one of their investigative tools.

Now to the actual story on how my involvement with water memory started and then kept unfolding. In the late 1980s the renowned British journal *Nature* published some reports that allegedly supported the claims broadly made by members of the alternative medicine community. These reports came from a group headed by a French immunologist, Jacques Benveniste (1935–2004), who was working in a reputable research institute in Paris. The most peculiar thing about these reports was that they seemed to provide validation for the practice of homeopathy—one of the most controversial areas of alternative medicine. In essence, homeopathy posits that when a certain drug is repeatedly diluted in water many times over (to the degree that not a single molecule of the original drug remains in the vessel, or, as chemical physicists would say, it is diluted 'below the Avogadro limit'), the water still somehow 'remembers' the action of this drug and hence it is still capable of delivering a healing action.

Such an inference (that water can have a memory) remains at odds with traditional chemistry and physics because these sciences (so far, at least) have been unable to produce any credible mechanism for such a memory in liquid substances. Likewise, mainstream medical professionals (with rare exceptions) do not recognize homeopathy as a valid medical practice and, hence, dismiss it as quackery and charlatanism. For example, no less an authority than Richard Dawkins, famous for his work in genetics (*The Selfish Gene*) and for the more recent fuss over his book *The God Delusion*, wrote that we should not 'be seduced by homeopaths and other quacks and charlatans, [who must] consistently be put out of business' (quoted from *This Will Make You Smarter*, edited by John Brockman, Harper Perennial, 2012). My question here is how can Dawkins (and many others) be so upfront certain that there is no physical mechanism for memory in liquid substances like water? Do we already know everything about physics, chemistry or how Nature works? For me such a bold inference sounds like at the very least grossly unwarranted arrogance.

However, the fact is that in spite of all the scientific negativity towards homeopathy, it remains a widespread practice, with millions of people using it and often claiming positive results. I personally became interested in this whole Benveniste story from the time it became public. Almost immediately after *Nature* published reports from the Benveniste group (in 1988), there was a deluge of articles and letters in the same journal, as well as in many other research and public outlets, that vehemently denied these 'outrageous' claims. At the same time, a small minority of scientists, the author of this book among them, took a more cautious attitude of the 'why not?' kind, a stance commonly know as 'what if'.

Indeed, *what if* water, in spite of being liquid, nonetheless has some physically based capacity to retain a 'memory' of its past and, specifically, can 'remember' what substances it was in contact with?

After all, there are some ideas around about 'water psychology', or water as a 'strange attractor' (Berezin 2012, 2015). Maybe, indeed, physics and/or chemistry (and, actually, for these matters 'chemistry' and 'physics' are almost the same thing) can come up with some plausible mechanism for such a memory effect in water? After all, there have been many claims in science which at first appeared odd and unbelievable, but later turned out to be genuine discoveries and became a part of mainstream science. Why could the same thing not happen with water memory?

Thus, in spite of numerous upfront denials from the mainstream science community that this (memory in water) 'cannot happen because it cannot' (they never give any cohesive and sustainable explanations as to why it cannot), there were here and there some voices of dissent. This is what, for example, the Nobel Prize physicist Brian Josephson of Cambridge University (the 'Josephson effect' in superconductivity and 'Josephson junctions') says in his letter to the *New Scientist* (Josephson 1997, p 66): ' Simple-minded analysis may suggest that water, being a fluid, cannot have a structure of the kind that such a picture [of water memory] would demand. But cases such as that of liquid crystals, which while flowing like an ordinary fluid can maintain an ordered structure over macroscopic distances, show the limitation of such way of thinking. There have not, to the best of my knowledge, been any refutations of homeopathy that remain valid after this particular point is taken into account.'

So, the key question here is how can isotopicity help water to 'remember its past'? Such a question, no matter how odd and esoteric it may seem, can (in the view of this author, at least) be scrutinized on the basis of microscopic atomic physics and the ideas of self-organization and informatics.

In science (and especially in physics) the ground rule for any research is to look for all the thinkable (and, sometimes, even unthinkable) options and ideas that may explain observable effects. Or, this may even concern effects that are claimed to exist by a sufficient number of people (like flying saucers, or the recently announced 'cold fusion'). And whether the end results (although there can hardly be 'end results' in science) lead to the confirmation of the effect or its refutation (the latter is a really hard thing to do), or otherwise leave the issue wandering in the limbo of uncertainty— to put all the possible options on the table is still a worthy thing to do (at least, in my humble opinion).

So, what about water memory? Let us assume for a minute that it is a proven fact that water has a memory effect. Now, what might our guesses may be as to what possible physical basis this might have? Several guesses had been offered, e.g. in the book *Ultra High Dilution—Physiology and Physics*, which featured several authors, including myself (Berezin 1994c). Without dismissing any alternative or competing (or complementary?) explanations (e.g. Del Giudice and Pulselli 2010), I will follow my 'isotopicity track' here.

So, perhaps, there is something in the water (yes, in pure water) that makes it possible for it to have a memory akin to our own. If this assumption sounds too fantastic, let us recall how we humans store our memories and process our thoughts. We do it through the complicated chemistry and physiology of the neurons in our brains. And what are they? Quite involved structures made of polymer molecules which are capable of forming multiple microscopic connections and interactions. Information is stored in them in a way that is not unlike the information storage in computer hard drives or digital cameras. There are several types of memory (magnetic, optical, electrostatic, etc), but they all come down to digital 'bits and bytes'. Now, the natural question to ask is whether water has some underlying structure in it that makes it a system capable of holding bits and bytes.

An affirmative answer to this question can be sought at the atomic level. The idea that the present author suggested (Berezin 1990a, 1990b, 1994c, 2015) focuses on isotopic diversity in water. Water, as everybody knows, is $H_2O$, it consists of oxygen and hydrogen; chemically speaking, it is hydrogen oxide. But both hydrogen and oxygen are mixtures of stable isotopes. Hydrogen has two stable isotopes, H (normal hydrogen) and D (deuterium), while oxygen has three isotopes ($^{16}O$, $^{17}O$ and $^{18}O$). To restate, isotopes are atoms that have the same position in the periodic table (that is, they are the same chemical elements) but differ in mass. This is because they have a different number of neutrons in their nuclei.

For example, the nucleus of ordinary hydrogen has just one proton (positively charged particle), while a D (deuterium) nucleus consists of a proton *and* a neutron (a neutron has almost the same mass as a proton, but it carries no electrical charge). Likewise, the above three stable isotopes of oxygen all have eight protons each, but they differ in the number of neutrons they have (eight, nine and ten, respectively). Thus, due to their mass difference, isotopes of the same chemical element are distinguishable atoms and, hence, their different combinations can carry information. For example, the chain of $^{16}O$ and $^{17}O$ isotopes can, in fact, be interpreted as an information-carrying binary string, say, 011010001101001 ... (etc).

In a series of papers (Berezin 1990a, 1990b, 1994c) I suggested that this is indeed what may happen in water. Memory in water can be 'plugged in' and retained in the combinations of isotopes. In fact, this would be a case of isotopic information storage. But water is a liquid. So, is there any plausible physical mechanism to stabilize the possible information content in water against the motions of water molecules? Several possible options for that were proposed in my papers, such as electrostatics (Berezin 1995), polarizational effects (Berezin 1983) and the Anderson localization mechanism (Berezin 1982a, 1982b, 1984f, 1986a).

Theoretically at least, there are six isotopically different kinds ('brands'?) of water in which two isotopes of hydrogen (H and D) can be combined with three isotopes of oxygen (we are talking here of stable isotopes, of course). In addition to this, there are almost unlimited ('continuous') possibilities for the mixing of isotopes in any desired proportions to obtain all kinds of isotopically enriched water. While D is only 0.0156% of all the hydrogen atoms in water (one deuterium atom per 6420 hydrogen atoms), in absolute numbers it comes to very impressive figures. One small drop of water (say, one cubic millimeter) still contains some $10^{15}$ atoms of deuterium. Using artificial methods, D and H can be separated and the so-produced water (in which most hydrogen atoms are D) is known as heavy water ($D_2O$). Deuterium and hence heavy water are *not* radioactive, so you cannot contract radiation sickness if you bathe in it.

But what would happen if you accidently drank a glass of heavy water? Some experiments in which mice were fed with heavy water showed some adverse effects due to the slowing of the metabolism. Although this author is unaware of any such tests involving human subjects, he would not recommend drinking heavy water (bottled or not), even if (contrary to what many people wrongly believe) heavy water is not radioactive. Deuterium is a stable isotope. However, in the general spirit of the tenets of homeopathy (in which highly diluted substances can be medically active), some small departures from the standard isotopic abundance may, perhaps, have some therapeutic effects (isotopic drugs?). The author is unaware of any such studies in the health sciences field.

To repeat what was said in the introduction and early sections of this book, many areas and aspects of stable isotopicity are considered here that have been discussed by this author in previous publications. Can isotopicity be used, for example, to build a new type of random number generator (Berezin 1987b), or a new type of isotopic optical fiber (Berezin 1988, 1989a, 1989b, 1992b), or can it be used to make microchips for quantum computers (Berezin 2009)? And what about compact information storage (Berezin 1984i)? Do isotopes affect brain function? Or, perhaps, could isotopes be an essential aspect of the very mechanism of consciousness (Berezin 1987c, 1992a, 1994a, 1994b)? Could there be an 'isotopic life form' out there in the cosmos, as an alternative to the 'regular' chemistry-based biology (Berezin 1984g, 1986b, 1987a, 1988e, 1990b, 2015)? Can life be based on a single chemical element (Berezin 1984e)? In addition to this, some philosophical and metaphysical discussions appear here and there in this book.

As my prior experience indicates, some people (who bother to read this book— and I do not assume that there will be any readers at all) may likely take the view that much, if not the majority, of what is presented in this book should be classified as fringe science, pseudo-science, or whatnot. Some of these ideas may challenge the established mainstream scientific orthodoxy, as many radical ideas have before. Many such ideas are indeed dead-end offerings, yet some 'off-line' ideas have at times turned out to bear fruit, and not always of the kind the original authors expected or intended.

To that effect, I always was, and remain, on the side of those who believe that unusual and off-mainstream ideas and suggestions should be heard and studied rather than dismissed outright as 'rubbish', as some critics are rushing to suggest (e.g. Rousseau 1992). The history of science offers many confirmations of this. It is abundant with examples of premature dismissals and the ridiculing of unorthodox ideas because they 'did not fit', as in Lavoisier's 'no stones can ever fall form the sky' quoted above. Even abstract mathematical ideas have a history of rejection and opposition. It is sufficient to mention the resistance to the idea of 'zero' (Seife 2000) or Cantor's theory of infinite sets (Dauben 1977, 1979).

Now, I would like to bend my story in a somewhat different direction. Fantastic as it may seem, I want to talk about a seemingly totally unrelated household item— the Möbius strip and its story (section 6.4). The next few lines explain the reason for such a departure from my isotopic narrative.

To my best knowledge at the time of writing (February 2016), the above ideas regarding isotopic information storage in living and quasi-biological systems (if we grant water the status of (semi)-living substance) have not been picked up by the research community or followed up through targeted experiments (in spite of the fact that I suggested some experiments in my papers). This prompted me to include two of my earlier essays in this book, which, in this context, appear to be quite relevant to the situation (see sections 6.4 and 6.5)—even though on the surface of things it may appear that these are from an entirely different opera. What does the Möbius strip (see below) have to do with water memory and homeopathy? Well, just wait until chapter 6 and keep on reading.

So, what can these ideas do for a controversial area of alternative medicine such as homeopathy? In this section and the next one, I will discuss the possible role of isotopic diversity in providing a rational basis for two widely claimed phenomena— the homeopathic effect and so-called 'crystal healing'. Both of these practices are commonly associated in the public perception with the notion of 'holistic medicine' and both are often debunked outright as unquestionable nonsense and charlatanism.

In brief, homeopathic action is defined as the ability of some specific drugs to retain their activity even after a number of multi-staged dilutions in water. Some activity has been claimed even for dilutions where virtually not a single molecule of the original drug remains in the container. As was just mentioned, in 1988 a widely publicized controversy occurred in connection with the so-called Benveniste affair— experiments with ultra-diluted water solutions of drugs. Although the experiments showed poor reproducibility and were severely criticized, the sources of the presumed errors (whether the whole phenomenon is an artifact or 'real'?) were never completely identified. The whole history of homeopathy is, of course, not restricted to this one particular experimental claim and, therefore, it seems justified to discuss the physical models for the presumed homeopathic effect as such, regardless of the value of Benveniste's experiments.

The physical model of the homeopathic effect should explain, at least hypothetically, how the water matrix could 'template' information on the nature and/or type of action of the primary (seed) drug molecules and then 'broadcast' this information down to subsequent dilution stages. In the case of the $H_2O$ matrix there are three isotopic degrees of freedom (H to D and $^{17}O$ or $^{18}O$ to $^{16}O$) and the concentration of the minority isotopes is not at all negligible: e.g. for $^{18}O$ it is 0.2% (one per 500), which corresponds to only eight lattice spacings $((500)^{1/3} = \sim8)$ for the average separation between two neighboring $^{18}O$ atoms. For comparison, 1/500 is an enormously high impurity concentration in doped materials (such as p- or n-doped silicon in computer electronics). Therefore, although the effects related to isotopic diversity are generally considered to be weak by common chemical standards, the ubiquity of isotopic diversity at practically every micro-spot might lead to some non-trivial consequences. In the stream of our isotopic paradigm, it is possible to suggest that the inherent isotopic diversity of water is at work, and that some positional correlations of stable isotopes (H, D, $^{16}O$, $^{17}O$ and $^{18}O$) might work as 'templates' of the originally dissolved molecules. The induction of isotopic correlations is equivalent to the choice of a particular isotopic pattern out of the highly degenerate manifold of potential patterns. This process bears some similarity to the reduction of the wave function of a particle (or a more complex system) during quantum mechanical measurement.

The very possibility of different positional organizations for minority isotopes (D, $^{17}O$, $^{18}O$) within the main $H_2O$ matrix leads to an enormous degree of 'isotopic redundancy' for the potentially available isotopic patterns. The essence of the proposed physical model for Benveniste's observations is that the presence of certain molecules (e.g. antibodies) might produce some specific readjustments in the positional distribution of minority isotopes in the vicinity of a given molecule. There might be several plausible ways in which this selection could work for the process of isotopic ordering, i.e. how the information on the nature of a dissolved molecule could be templated into a positional arrangement of isotopes. For example, it is known that ionic polarizability is mass-dependent (the vibrational frequency is proportional to $M^{(-1/2)}$; $M$—atomic mass) and should therefore be rather sensitive isotopically.

Some positional combinations of isotopes (e.g. clusters of $^{18}O$) could enhance the local values of the polarizability of the media. The deepening of the polarization

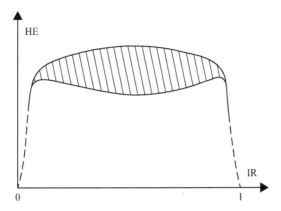

**Figure 5.1.** Homeopathic efficacy (HE), as an expected function of a particular isotopic ratio (IR). The existence of a long 'plateau' is a likely consequence of a high redundancy of isotopic information storage. Due to the uncertainties in 'quantifying' of 'HE' the 'plateau' region area is shown as a shadowed area of some (arbitrary) thickness.

potential wells may serve as a stabilizing factor in a similar way to the polaronic self-stabilization in crystals (Berezin 1983). This may explain the possible robustness of isotopic correlations against disordering thermal effects and also account for the reduplication ('Xeroxing') of isotopic correlations at sequential dilution steps (figure 5.1). Moreover, spatially correlated isotopic arrangements could eventually explain the oscillation effects in Benveniste's experiments (the bioactivity shows an oscillating rather than a monotonic dependence on the dilution level). The clue to this could lie in an analogy with the commensurate–incommensurate transitions in partially ordered isotopic superlattices (Berezin 1988a, 1988c, 1988b, 1989b, 1990a, 1990c, 1991b).

A similar scenario could lead to a variable (oscillating) degree of isotopic ordering in the water matrix. The mechanism of such transformations is often referred to in solid-state physics under the term 'devil's staircases'. It acts in some specific crystalline (so-called Ising) systems with lattice frustration. The latter results in oscillating patterns of crystal ordering with a continuous change of the coupling constant or concentration ratio.

These are my tentative guidelines for a possible rational explanation of the 'unbelievable' effect reported by Benveniste and others. It is based on the concept of isotopic patterns and the ability of isotopically organized structures to store, transfer and, perhaps, even amplify information. This explanation does not dismiss the possibility of alternative explanations, as discussed in a variety of sources. Moreover, it could turn out to be complementary to them. I should also note that the above hypothesis offers a line of experimentation, since isotopic ratios are relatively easy to change artificially (e.g. through increasing the concentration of the $^{18}O$ isotope in water) and the described isotopic effects could therefore be rather easily enhanced or suppressed in terms of their role in these processes.

## 5.17 Isotopic neural networks

At the atomic level it can easily be seen that in isotopically mixed crystal lattices there could be a number of similarly structured isotopic micro-complexes. This might lead to low-frequency vibrational resonance effects and account for memory storage phenomena. For example, holographic-type memory effects in quartz crystals may be related to complexes that involve minority isotopes of oxygen and/or silicon (e.g. $^{17}O$ and $^{29}Si$) and they could be describable as the formation of interactive connections that we shall provisionally call 'isotopic neural networks'. They could operate in a manner similar to the known neural networks in spin glasses (Hopfield 1982). Such non-linear interactive systems are capable of spontaneous self-organization in the sense of developing highly correlated patterns of site states in time and space, showing 'behavioral patterns' of chaos and organized activity and imitating evolutionary processes such as self-complication.

Quartz is often claimed to be the most popular of the alleged healing crystals. It is interesting to note that quartz has more than one isotopic sub-lattice. This provides a basis for the possibility of a 'natural division' of functions in which one isotopic sub-lattice (e.g. $^{17}O$) plays the role of neural sites (two possible spin states), while the other (e.g. $^{29}Si$) plays the role of a synaptic network. An example of such an isotopic neural network is shown in figure 5.2.

The 'healing action' in this context could eventually mean that the informational interaction between the human body and/or mind ('consciousness') and a crystal 'locks' the latter in a state in which a particular physiological or psycho-logical pattern is templated into the crystal. This templated pattern can, in turn, provide informational feedback to support the originally chosen modality of the physiological or psychological state. In other words, the isotopic degree of freedom could provide a basis for the establishment of an informationally interactive linkage between a crystal and a human body. Polyisotopic neural networks probably possess a high(er) degree of information storage redundancy and, therefore, they are likely

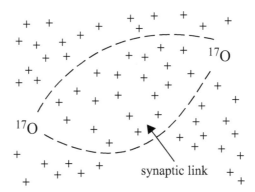

**Figure 5.2.** Isotopic diversity in crystalline quartz (SiO₂) presented as a basis for an 'isotopic neural network'. Only isotopes with non-zero nuclear spins and magnetic moments ($^{29}Si$: crosses) and $^{17}O$ are shown. Area circled by dashed line indicates an approximate region of 'synaptic link' of a few dozens of 29Si atoms connecting two 17O atoms.

to be quite robust in terms of pattern storage and resistance against disordering factors.

As the above discourse points out, the informational aspects of isotopicity can be advanced in terms of the currently important paradigm of neural networks. As was mentioned in a previous chapter, in isotopically mixed crystal lattices there could be a number of similarly structured isotopic micro-complexes. This might lead to low-frequency vibrational resonance effects and account for the memory storage phenomena which are strongly advocated by some adherents and practitioners of so-called crystal healing. Many people share the feeling that crystals are somehow (almost) 'living beings', that they are some alternative life forms with whom people can establish personal connections. These views have a long history and tradition in many cultures and are connected to a number of mythological and esoteric teachings and metaphysical systems. The recent so-called New Age movement revived the popularity of these ancient views and practices in a modern context.

Without discussing, or in any way critically assessing, every aspect of New Age claims in this book, I would still like to indicate that isotopicity could provide (at least hypothetically) a physical model for some of the effects claimed in connection with healing crystals (Berezin 1991a). I certainly leave it up to the taste of individual readers to look for further arguments to support or refute the hypothesis that is outlined here. Although, in my personal view, it would not be much easier to refute these claims once and for all than it would be to refute 'convincingly' the efficacy of placebos, or, for that matter, to refute the existence of parallel Universes or any other 'transcendental reality', or, indeed, to refute the notion that we all may be living in a simulated reality, in other words, that we may be virtual beings simulated in some super-computer (Berezin 2006 and section 6.11 below).

Let us consider the isotopic diversity of the most popular of the healing crystals, namely quartz. As a silicon dioxide ($SiO_2$), quartz has as many as six isotopic sub-lattices (both oxygen and silicon have three stable isotopes each). The alleged holographic-type memory effects in quartz crystals could be related to complexes that involve minority isotopes of oxygen and/or silicon (e.g. magnetically active $^{17}O$ and $^{29}Si$ isotopes) in a manner describable as the spontaneous formation of 'isotopic neural networks', which would be similar to the neural networks suggested for spin glasses (Hopfield 1982).

To restate, the fact that most crystals have more than one isotopic sub-lattice provides the basis for a natural division of functions, with one isotopic sub-lattice (e.g. $^{17}O$) playing the role of neural sites (two possible spin states), while the other (e.g. $^{29}Si$) plays the role of the synaptic network. In a quartz crystal there are about 60 atoms of $^{29}Si$ per single atom of $^{17}O$. One can see the whole 'community' of those $^{29}Si$ atoms (say, 20 or 30) that are located between any two neighboring $^{17}O$ atoms as providing a synaptic link between the latter. Such a $^{29}Si$ cluster has many inner spin states, i.e. each of the 20 or 30 atoms may have two nuclear spin directions, so there are many possible combinations.

As a result, the so-designated $^{29}Si$ sub-network has the potential for almost continuous synaptic adjustments. As far as the virtually boundless informational capacity of isotopically diversified crystal is postulated, the problem of interaction between the crystal and the host (e.g. human body) becomes relatively secondary since such an informational exchange could be attained by a whole number of means (e.g. through resonance at infra-sound frequencies or through magnetic or electro-static effects, etc). To conclude the above argument, it should be pointed out that this whole hypothesis could, in principle, be subjected to experimental tests, e.g. using crystals with different isotopic compositions.

However, there may be another 'cosmic' angle to these ideas. In the context of electrostatic self-organization (Berezin 1994d), it is interesting to note the recently discovered DNA-like plasma structures. An important recent development in the area of dusty plasma self-organization was the discovery of the formation of helical spiral structures. Experiments performed in the weightless (microgravity) conditions at the International Space Station (ISS), as well as theoretical simulations, revealed this amazing phenomenon occurring in charged dusty plasmas. They showed a propensity to spontaneously self-organize themselves into helical (spiral) structures that resemble double DNA spirals (Nefedov *et al* 2003, Melzer 2006, Tsytovich *et al* 2007, Kamimura and Ishihara 2012, Hyde *et al* 2013).

As the above-quoted investigations showed, complex plasmas may naturally self-organize themselves into stable interacting helical structures that exhibit features normally attributed to organic living matter and in particular DNA double spirals. These interacting complex structures exhibit thermodynamic and evolutionary features that are thought to be peculiar to living matter, such as bifurcations that serve as 'memory marks', self-duplication, metabolic rates in a thermodynamically open system and non-Hamiltonian dynamics. Likewise, these structures reveal faster evolution rates by competing for 'food' (surrounding plasma fluxes). In other words, these structures could have all the necessary features to form 'inorganic life'. Thus, we could be facing the fascinating possibility that inorganic life 'invents' organic life (Tsytovich *et al* 2007).

The far-reaching hypotheses coming out of this plasma work may point to the possibility of alternative (perhaps, non-carbon) forms of life, 'plasma life in the cosmos', in addition to providing some alternative scenarios for the origins of life on Earth (and perhaps on other planets). In view of the ideas on the possible role of isotopicity (isotopic diversity) presented in this book ('isotopic biology'), it seems tempting to propose that isotopic effects may, in turn, lead to some non-trivial effects in these 'plasmodic structures'. At the same time, I realize that speculating further on these possibilities in more concrete terms at this stage may be tantamount to science fiction along similar lines to such masterpieces as Lem's *Solaris* (Lem 1981). However, as is well known (there are many examples), science fiction more often than not has predicted the things that sooner or later become a reality. And sometimes the actual reality may even outdo the predictions of science fiction— many examples of this can be pointed out. So, 'isotopicity in plasma DNA' remains an open question until further data can be obtained and analyzed.

However, looking at the ISS experiments on dust plasma self-organization in microgravity conditions (Nefedov *et al* 2003), which are supposed to imitate similar effects in the open cosmos, we can observe the following. Taking 5 μm as a typical size for dust particles (Nefedov *et al* 2003, Tsytovich *et al* 2007), we can estimate that such a particle typically contains a trillion ($10^{12}$) or so atoms. Such a number could provide ample opportunity for all sorts of isotopic arrangements. Hence, the arguments about the possibility of the formation of isotopic neural networks presented in this book and in my earlier publications (Berezin 1990b, 1991a, 1992b) may well be applicable to such dust particles. If so, such an 'isotopic enrichment' of the situation with 'plasma DNA in the cosmos' could greatly enhance the information processing and information storage capacity of such structures.

# References

Abraham R H 1994 *Chaos, Gaia, Eros. A Chaos Pioneer Uncovers the Three Great Streams of History* (San Francisco, CA: Harper)

Aczel A D 2000 *The Mystery of the Aleph: Mathematics, the Kabbalach, and the Search for Infinity* (New York: Four Walls Eight Windows)

Allahverdyan A E and Nieuwenhuizen T M 2006 Explanation of the Gibbs paradox within the framework of quantum thermodynamics *Phys. Rev.* E **73** 066119

Anderson P W 1958 Absence of diffusion in certain random lattices *Phys. Rev.* **109** 1492–505

Arnold A 1992 *The Corrupted Sciences. Challenging the Myths of Modern Science* (London: Paladin)

Bellavite P and Signorini A 2002 *The Emerging Science of Homeopathy: Complexity, Biodynamics and Nanopharmacology* (Berkeley, CA: North Atlantic)

Berezin A A 1969 Theory of the Auger effect in a system of two crystal defects *Fizika Tverdogo Tela* **11** 1587–90 (in Russian)

Berezin A A 1969 Theory of the Auger effect in a system of two crystal defects *Sov. Phys.—Solid State* **11** 1285–7 (Engl. transl.)

Berezin A A 1970 Theory of positron annihilation at F-centers *Fizika Tverdogo Tela* **12** 3315–7 (in Russian)

Berezin A A 1971 Theory of positron annihilation at F-centers *Sov. Phys.—Solid State* **12** 2684–5

Berezin A A 1982a Anderson localization induced by an isotopic disorder *Lett. Nuovo Cimento* **34** 93–6

Berezin A A 1982b On the Anderson transition in electronic color centers systems in alkali halide crystals *Z. Nat. forsch.* A **37** 613–4

Berezin A A 1983 Spontaneous tunnel transitions induced by redistribution of trapped electrons over impurity centers *Z. Nat. forsch.* A **38** 959–62

Berezin A A 1984a Excited states deactivating exclusively through correlated electron tunneling *J. Chem. Phys.* **81** 851–4

Berezin A A 1984b Instabilities against correlated two-electron tunnel transitions in impurity systems *Solid State Comm.* **49** 87–9

Berezin A A 1984c Two-electron-one-photon transitions in polycenter systems *Chem. Phys. Lett.* **104** 226–8

Berezin A A 1984d Double tunnel jumps in activationless hopping conductivity *J. Phys. C: Solid State Phys.* **17** L393–7

Berezin A A 1984e Can life be based on a single chemical substance? *Naturwissenschaften* **71** 45 (Proposed (speculative) idea that the isotopic diversity in such simple compounds as water can lead to information bearing structures and be a foundation of alternative quasi-biological activity.)

Berezin A A 1984f An isotopic disorder as a possible cause of the intrinsic electronic localization in some materials with narrow electronic bands *J. Chem. Phys.* **80** 1241–5

Berezin A A 1984g Isotopic biology *Nuovo Cimento* D **3** 914–6 (Discussion of the possibility of alternative biology based on isotopic combinations.)

Berezin A A 1984h Information storage based on isotopic combinations *Specul. Sci. Technol.* **7** 317–9

Berezin A A 1984i Inherent genetic instability due to the presence of $^{14}$C atoms in information bearing structures *Bull. Am. Phys. Soc.* **29** 635

Berezin A A and Jamroz E J 1984 Collectively deactivating excited configurations of impurity systems in electric field—electroluminescence and switching *J. Lumin.* **31/32** 188–90

Berezin A A 1986a Quasi-localization and electronic transport in boron and boron-rich borides *International Conference on the Physics and Chemistry of Boron and Boron-Rich Solids (Albuquerque, NM, 1985)* vol 140 (New York: AIP) pp 224–33

Berezin A A 1986b On the mechanisms of information transfer in isotopic biology *Kybernetes* **15** 15–8

Berezin A A 1987a Stable isotope engineering in life studies biotechnology and bioengineering **30** 798

Berezin A A 1987b Isotopic jets as a perfect random number generator *Int. J. Electron.* **63** 673–5

Berezin A A 1987c Isotopic dimension in biology—a counterpart for quantum indeterminism? *Nuovo Cimento* D **9** 731–3

Berezin A A 1987d Super super large numbers *J. Recreat. Math.* **19** 142–3

Berezin A A 1988a Isotopic superlattices and isotopically ordered structures *Solid State Comm.* **65** 819–21

Berezin A A 1988b Light bending and light confinement at isotopic boundaries—possibility of isotopic fiber optics *Thin Solid Films* **158** L37–8

Berezin A A 1988c Isotopic ordering and isotopic correlations as a possible new tool for geosciences *Chem. Geol. (Isot. Geosci. Sect.)* **72** 197–8

Berezin A A 1988d Isotopic randomness—some fundamental and applied aspects *Phys. Essays* **1** 133–7

Berezin A A 1988e Isotopic randomness as a biological factor *Biol. J. Linnean Soc. (London)* **35** 199–203

Berezin A A and Ibrahim A M 1988 Effects of the diversity of stable isotopes on properties of materials *Mat. Chem. Phys.* **19** 407–30

Berezin A A 1989a Isotopic engineering (perspectives) *J. Phys. Chem. Solids* **50** 5–8

Berezin A A 1989b Some effects of positional correlations of stable isotopes *Phys. Lett.* A **138** 43–5

Berezin A A 1990a Isotopical positional correlations as a possible model for Benveniste experiments *Med. Hypotheses* **31** 43–5

Berezin A A 1990b Isotopic diversity as an unexplored mind–matter dimension *Sci. Prog.* **74** 495–512

Berezin A A 1990c News on quantum foundations of consciousness *Perceptual Motor Skills* **70** 930

Berezin A A and Nakhmanson R S 1990 Quantum mechanical indeterminism as a possible manifestation of microparticle intelligence *Phys. Essays* **3** 331–9

Berezin A A 1991a On the possible physical foundations of health related effects of crystals *Med. Hypotheses* **36** 213–5

Berezin A A 1991b Isotopic relatives of strange attractors *Phys. Lett.* A **161** 295–300

Berezin A A and Ibrahim A M 1991 Semiempirical model of isotopic shifts of the band gap *Phys. Rev.* B **43** 9259–61

Berezin A A 1992a Correlated isotopic tunneling as a possible model for consciousness *J. Theor. Biol.* **154** 415–20

Berezin A A 1992b Isotopicity: implications and applications *Interdiscip. Sci. Rev.* **17** 74–80

Berezin A A 1994a Isotopes, information and quantum self *Frontier Perspect* **4** 36–8

Berezin A A 1994b The problem of ultimate reality and meaning in the context of information self-organization and isotopic diversity *Ultimate Real. Meaning* **17** 295–309

Berezin A A 1994c Ultra high dilution effect and isotopic self-organisation *Ultra High Dilution. Physiology and Physics* ed P C Endler and J Schulte (Dordrecht: Kluwer) pp 137–69

Berezin A A 1994d Quantum aspects of self-organization in dynamically random systems *Dusty and Dirty Plasmas, Noise, and Chaos in Space and in the Laboratory* ed H Kikuchi (New York: Plenum) 225–40

Berezin A A 1995 Electrification of solid materials *Handbook of Electrostatics* ed J S Chang, A J Kelly and J M Crowley (New York: Marcel Dekker) pp 25–38

Berezin A A 1996 Mainstream and fringe scientific ideas and ultimate values *Ultimate Real. Mean.* **19** 40–9

Berezin A A 1998 Meaning as self-organization of ultimate reality *Ultimate Real. Meaning* **21** 122–34

Berezin A A 1999 Distant life *Discover* **20** (December 1999) pp 20–21

Berezin A A 2002 Energy, information, and emergence in the context of ultimate reality and meaning *Ultimate Real. Meaning* **25** 256–73

Berezin A A 2004 Ideas of multidimensional time, parallel universes and eternity in physics and metaphysics *Ultimate Real. Meaning* **27** 288–14

Berezin A A 2006 Simulation argument in the context of ultimate reality and meaning *Ultimate Real. Meaning* **29** 244–61

Berezin A A 2009 Stable isotopes in nanotechnology *Nanotechnol. Perceptions* **5** 27–36

Berezin A A 2011 Isotopic engineering in surface science and technology *Surface Effects and Contact Mechanics X, 10th International Conference on Surface Effects and Contact Mechanics (Malta, September 2011)* ed J T M De Hosson and C A Brebbia (Southampton: WIT Press) pp 193–204

Berezin A A 2012 Aquatic realms and running water in sustainable tourism *Environmental Impact* ed C A Brebbia and T-S Chon (Southampton: WIT Press) pp 447–58

Berezin A A 2015 *Isotopicity Paradigm: Isotopic Randomness in the Digital Universe* (Cambridge: Cambridge International Science)

Comorosan S 1974 The measurement problem in biology *Int. J. Quantum Chem. Biol. Symp.* **1** 221–8

Condorcet C M J A N 1795 1955 *Sketch for a Historical Picture of the Progress of the Human Mind* (New York: Noonday) (Engl. transl.)

Dauben J W 1977 Georg Cantor and Pope Leo XIII: mathematics, theology, and the infinite *J. Hist. Ideas* **38** 85–108

Dauben J W 1979 *Georg Cantor: His Mathematics and Philosophy of Infinite* (Princeton, NJ: Princeton University Press)

Del Giudice E and Pulselli R M 2010 Formation of dissipative structures in liquid water *Int. J. Design Nat. Ecodynamics* **5** 21–6

Deutsch D 1997 *The Fabric of Reality* (London: Allen Lane)

Eccles J C 1986 Do mental events cause neural events analogously to the probability fields of quantum mechanics? *Proc. Royal Soc. London* **227** 411–28

Epling G A and Florio E 1981 Isotope enrichment by photolysis on ordered surfaces *J. Am. Chem. Soc.* **103** 1237–8

Grib A A and Rodrigues W A 1999 *Nonlocality in Quantum Physics* (New York: Kluwer)

Hopfield J J 1982 Neural networks and physical systems with emergent collective computational abilities *Proc. Natl Acad. Sci. USA* **79** 2554–8

Hyde T W, Kong J and Matthews L S 2013 Helical structures in vertically aligned dust particle chains in a complex plasma *Phys. Rev. E* **87** 053106

Ibrahim A M and Berezin A A 1992 Synthesis of buried insulating layers in silicon by ion implantation *Mater. Chem. Phys.* **31** 285–300

Jahn R G and Dunne B J 1988 *Margins of Reality: The Role of Consciousness in the Physical World* (New York: Harcourt Brace Jovanovich)

Jaynes E T 1992 The Gibbs paradox *Maximum Entropy and Bayesian Methods* ed C R Smith, G J Erickson and P O Neudorfer (Dordrecht: Kluwer) pp 1–22

Josephson B 1997 *New Scientist* (1 November 1997) p 66)

Kafatos M and Nadeau R 1990 *The Conscious Universe: Part and Whole in Modern Physical Theory* (New York: Springer)

Kamimura T and Ishihara O 2012 Coulomb double helical structure *Phys. Rev. E* **85** 016406

Katz J J 1960 Chemical and biological studies with deuterium *Am. Sci.* **48** 544–80

Keswani G H 1986 Carbon-14 in the brain and other organs: chance within *Specul. Sci. Technol.* **9** 243–4

Knuth D E 1976 Mathematics and computer science: coping with finiteness *Science* **194** 4271

Kurzweil R 2005 *The Singularity is Near: When Humans Transcend Biology* (New York: Viking)

Lanza R and Berman R 2009 *Biocentrism: How Life and Consciousness Are the Keys to Understanding the True Nature of the Universe* (Dallas, TX: BenBella)

Lem S 1981 *Solaris* (New York: Penguin)

Lloyd S 2006 *Programming the Universe: A Quantum Computer Scientist Takes on the Cosmos* (New York: Knopf)

Mann A K and Primakoff H 1981 Chirality of electrons from beta-decay and the left-handed asymmetry of proteins *Orig. Life* **1** 255–65

McTaggart L 2003 *The Field: The Quest for the Secret Force of the Universe* (New York: Harper Collins)

Melzer A 2006 Zigzag transition of finite dust clusters *Phys. Rev. E* **73** 056404

Nefedov A P *et al* 2003 Plasma crystal experiments on the International Space Station *New J. Phys.* **5** 33

Oreskes N 1999 *The Rejection of Continental Drift* (Oxford: Oxford University Press)

Penrose R 1994 *Shadows of the Mind* (Oxford: Oxford University Press)

Penrose R 1996 On gravity's role in quantum state reduction *Gen. Relativ. Gravit.* **28** 581–600

Popper K and Eccles J C 1977 *The Self and Its Brain* (Berlin: Springer)

Pui J P and Berezin A A 2001 Mind, matter, and diversity of stable isotopes *J. Sci. Explor.* **15** 223–8

Rousseau D L 1992 Case studies in pathological science *Am. Sci.* **80** 54–63

Sams G 2009 *Sun of gOd* (San Francisco, CA: Weiser)

Schrödinger E 1945 *What is Life?* (Cambridge: Cambridge University Press) (also numerous subsequent editions)

Seife C 2000 *Zero: The Biography of a Dangerous Idea* (New York: Penguin)

Siler T 1990 *Breaking the Mind Barrier: The Artscience of Neurocosmology* (New York: Touchstone)

Skewes S 1933 On the difference between $\pi$ (x)-Li(x)' *J. London Math. Soc.* **8** 277–83

Skewes S 1955 On the difference between $\pi$ (x)-Li(x) II *Proc. London Math. Soc.* **5** 48–70

Talbot M 1992 *Mysticism and the New Physics* (London: Arkana)

Thode H G 1980 Sulphur isotope ratios in late and early Precambrian sediments and their implications regarding early environments and early life *Orig. Life* **10** 127–36

Tsytovich V N, Morfill G E, Fortov V E, Gusein-Zade N G, Klumov B A and Vladimirov S V 2007 From plasma crystals and helical structures towards inorganic living matter *New J. Phys.* **9** 263

Zajonc A G 2003 Light reconsidered *Opt. Photonics News* **3** 2–5

Zohar D 1990 *The Quantum Self: Human Nature and Consciousness Defined by the New Physics* (New York: William Morrow)

**IOP** Publishing

Digital Informatics and Isotopic Biology

Self-organization and isotopically diverse systems in physics, biology and technology

**Alexander Berezin**

# Chapter 6

# Discovery and innovation in our digital society

Monographs in a particular research area—be they from physics, biology, medicine, engineering (etc)—do not *as a rule* have special extended chapters on the sociological, historical or economic aspects of the area in question. Of course, here and there we can find exceptions to this rule and the topic of isotopicity (the main subject of this book) provides good reasons to make such an exception. In this, the last, chapter of the book (before the conclusion) I discuss a variety of reasons why, as I maintain, the issue of stable isotopicity remains underappreciated and underused in its possible applications. This is especially so for the informational (digital) aspects of isotopicity in biology and biomedical research—the areas that are discussed in chapters 4 and 5. To better assess the whole situation regarding isotopicity, the sections of this chapter suggest that it be looked at from several perspectives.

## 6.1 The blessings and evils of global interconnectedness and digitization

*Civilization advances by extending the number of important operations which we can perform without thinking of them.*

Alfred North Whitehead (1861–1947)

The central theme of this book is the digital aspects of isotopic diversity and a variety of proposed implications of this fact (isotopicity) in several areas of science and technology. In this regard the very *zeitgeist of digitization* calls for a few words regarding our so-called 'digital age'. Like almost any watershed in history, the digital age that we are now experiencing brings with it a mixture of blessings and challenges. Furthermore, almost any aspect of the digital age often invokes diametrically opposed emotions in different people (there are many books on these dilemmas).

doi:10.1088/978-0-7503-1293-6ch6

It is an almost universally accepted premise that in our high-tech age the pace of scientific research and technological development is accelerating exponentially. This inference appears to be so obvious that it would be tantamount to blasphemy to express much doubt about it. Unlike a political revolution, the digital revolution that we are witnessing before our eyes is not an event localized over a relatively short time period (as the French or Russian revolutions were), but is an ongoing process that is unfolding over several decades (Schmidt and Cohen 2013). It most likely will continue with, perhaps, exponential acceleration (Moore's Law, etc). So, it is a kind of 'permanent revolution', to use the notorious terminology of Leon Trotsky. This digital revolution is a play with so many participating actors (scientific, techno-logical, social, political and economic) that it is close to impossible to assign any specific 'starting date' for it. Was it the invention of the transistor, or the digital computer, or the internet and world wide web, etc—all such events (and many others) can be counted as milestones on the digital highway.

And yet, some recent skeptical overviews have exposes serious weaknesses in such an overtly optimistic scenario. Thus, under critical scrutiny, the question of whether we are moving too quickly, or not quickly enough, turns out to be not so straightforward as it may appear at first glance. In the unstoppable tsunami of new consumer goods falling on us daily, it is becoming more and more difficult to distinguish between real innovation and mere market repackaging of older products. The same effect (the growing difficulty in distinguishing between the genuinely new and the recycled) is also apparent in the realm of scientific and technological ideas.

The tsunami of informational devices that has fallen on us in recent decades affects us and our lives in many ways. From small personal devices to the global internet, these things are accelerating almost exponentially. And the process still keeps up its pace, with no signs of abating. Moore's law (in which processing power doubles every two years—or, some say, every 18 months) is still doing its job without taking any time off. And all this brings about inevitable adjustments to our very personalities. In a number of key ways we are acquiring 'digital personalities' and there is not even much need for me to discuss this process at length—the evidence for this is abundant for anyone who takes a moment to look around and contemplate. And this process is taking its toll on the dynamics and difficulties of new discoveries and innovation.

## 6.2 The paths and mazes of science and discovery

*It is necessary to be slightly underemployed if you are to do something significant.*

James D Watson (b 1928), co-discoverer of the DNA genetic code

The boundaries between the mainstream and the fringe are to some degree arbitrary, or at least fuzzy. Traditionally, the notion of mainstream science includes the main contents of the major natural sciences, such as physics, chemistry, biology, geology and most engineering-related disciplines.

This definition, however, backfires with the effect that almost anything can be exempt. For example, (pure) mathematics is not seen by some people as being 'science' in the proper sense (there are no 'experiments' in pure mathematics) and is instead viewed as a philosophy of the eternal and immutable (Platonic) world (Dauben (1977), (1979), Lavine (1994), Penrose (1996), Tegmark (2014)). Likewise, many other areas which traditionally count as sciences cannot be fully 'objective' for almost the opposite reason—they often carry too heavy an ethical, social and sometimes political load, e.g. such areas as psychology or some aspects of the medical sciences.

Hence, almost any commonly held perception about what constitutes mainstream science requires some individual specification. For the purpose of this discourse, let us adopt the following working definition of mainstream science. We can define it as a body of knowledge that is built up on the questions 'how?' (e.g. how does 'it' work, how has 'it' evolved or how is 'it' evolving, etc) and 'what is it?' (e.g. what is an isotope?).

Within this framework the question 'why?' is meaningless, or at least has no convenient placement except in a cause-and-effect chain. This is in spite of the fact that 'why?' is probably the most frequently asked question in our everyday lives. However, in a science such as physics it can only be properly asked in a structural or rhetorical sense. We can ask 'what is temperature?' (the average kinetic energy of molecules), but it is (almost) impossible to answer the question 'why is there a temperature?'. In mathematics, it is an even more acute dilemma. We can ask all we want about the pattern of prime numbers (Ribenboim 1989), but the question 'why is 17 a prime number?' is meaningless (if it calls for some explanation) and the only valid answer to it is 'because it is' (17 is a prime number because it is a prime number). It is like saying '2 + 2 = 4 on Sundays'—formally correct (who will object that it is wrong?), but a profoundly odd and meaningless sentence.

If it is defined as above, (mainstream) science can easily exist without any metaphysics-related traits, in the same way as commerce, manufacturing, construction and many other mass occupations. The alleged difference though is that scientists have to a large extent (though not exclusively) some genuine interest in the foundations of this world. Therefore, metaphysical connotations are not that rare even in mainstream scientific texts (though for the most part it is a game of escapism, akin to fishing, gardening or knitting). Here we come to another slice of it, another angle to look at this whole phenomenon of escapism, something which can be provisionally labeled the 'power of a ritual and/or the notion of a service' metaphor.

In developing the above theme it is useful to study the effect of 'ritualization emergence'. In many connotations the emergence of a ritual can be traced to the mythological and metaphysical aspects of the situation. In what way? The ritual implies the embedded repetition of some kind of a protocol (mantra, prayer or liturgy) for the purpose of transcending the restrictability of a given moment of time/space, to eternalize 'the present'.

We can observe that (almost) the entire realm of scientific activity can be considered under this angle. Let us again start with the example of pure mathematics.

Even a superficial exposure to the philosophical foundations of mathematics (e.g. Rucker 1987, Russell 1989, Lavine 1994) is likely to yield a conclusion that is a confidently robust message at the level of common sense. The message is simply this: what mathematics 'discovers' (literally: *dis*-covers) are 'in reality' some 'eternal' Platonic truths. Probably the best (and perhaps easiest to follow) example is provided by number theory (the theory of integer numbers).

The 'best' here means something that can reliably (robustly) appeal to a common sense. For instance, the intricacies of prime number distribution are forever fixed and unchangeable. Whatever we can 'dis-cover' about prime numbers (Dickson (1960), Arnold (1992), Plichta (1997), Pickover (1995), (2001)), is, of course, already 'known' to God (or a universal mind, or whatever metaphor is used in a particular metaphysical belief system). Thus, as all these truths have already have been discovered and rediscovered in the uncountably numerous worlds and baby Universes that are, so to speak, parallel worlds to our Universe (for some cosmological models the time sequentiality is unimportant, i.e., it does not matter if our world is 'before' or 'after' some other worlds—which are similar or not to ours).

There is an anonymous quotation that says 'Science is the game we play with God to find out what His rules are.' So, in this context the issue of a ritual or 'eternalized service' opens up as a kind of naturally intuitive development of the topic. In the so-called natural sciences (physics), dealing with the specifics of 'our' form of the reincarnation (or 'embodiment') of the laws of physics and the material Universe, the situation with ritualization at first glance seems to be less immediate. I suggest, however, that it is largely an illusion and the ritualization elements in physics are often as strong as in the 'much purer' world of mathematics.

Clashes in physics are on some occasions based on what are essentially pseudo-issues. Sometimes, such clashes can be resolved (or at least, mediated) by the admission of 'multiplicity of truth'-type philosophies (e.g. multi-valued logic, etc). An example is the controversy over the reproducibility of physical measurements. From the point of view of mainstream physics, reproducibility is an absolute must. No result in 'true physics' should bear any trace of the individuality of the discoverer. To admit such a possibility, according to mainstream thinking, is to betray the very foundations of the whole edifice of science. To hold the view that such 'objective physical parameters' as, say, the mass ratio of a proton and an electron ($M/m = 1836$) can depend (even in the 20th decimal digit) on the individuality of the experimentalist (or even more startlingly, on his or her 'intentions'), would normally be perceived as sheer lunacy. Such an assumption simply has no place within the framework of mainstream physical thinking.

One press-documented illustration of a controversy on the above issue can be seen in the polemics exchanged between two physicists working at the same university. Philip Anderson, professor of physics at Princeton University and a Nobel Prize winner (for the discovery, among other things, of the Anderson localization discussed elsewhere in this book), challenged Robert Jahn, head of the Princeton Engineering Anomaly Research Laboratory (Anderson 1990, 1991). Jahn's group had been studying the alleged effects of consciential intentions on the performance of electronic random number generators. The major claim made as a result of this work was that some operators do indeed have a statistically verified ability to 'bias' random number generators through mental effort (Jahn and Dunne 1988).

The possibility of such a 'spooky action' is anathema to mainstream thinking. Anderson explained why (Anderson 1990, 1991): 'If the "observer effect", as he [Jahn] calls it—or "magic", as one might equally well characterize it—is correct, precise measurement is not possible. His ideas are as incompatible with the intellectual basis of physics as "creation science" is with that of cosmology or biology'. Leaving aside the highly contentious issue of whether one should add extra weight to authoritativeness when a pronouncement is made by a Nobel Prize holder, I can indicate at least two open ends in the above (seemingly sensible) Anderson quotation.

1. Suppose that 'precise measurement' is indeed not possible and that the (very small) observer effect does exist—why should this augur the end of physics, as Anderson apparently fears?
2. The 'intellectual basis of science', as the historical record clearly indicates, is a rather fuzzy, controversial and ever-changing arena, rather than something we can agree upon once and for all.

However, what many philosophers and metaphysicians are searching for in all this feeble and ephemeral stuff is some kind of permanence, a solid foundation for our experience and its eternal validation. Physical science, despite its appearance of utter objectivity, has an irreducible contextual factor (Arnold 1992, Jahn and Dunne, 1988). Perhaps, we should indeed take more serious note of the famous quotation assigned to the German mathematician Leopold Kronecker (1823–91), 'God created only the integer numbers, all the others being the work of a man'.

What Kronecker apparently meant here is that the 'laws' of integer numbers (e.g. the distribution of prime numbers) are the only genuinely immutable and ever-the-same foundation of the world—even our (more esoteric) mental constructs, such as the infinite hierarchy of sets introduced by Georg Cantor (Dauben 1979) may bear the traces of contextual relativity. This may, perhaps, explain the animosity that Kronecker felt towards Cantor and his theories of infinite sets. Even such monumental constructs as Gödel's incompleteness theorem (which posits that the complete system of axioms always has unprovable statements in it) are challenged (Good 1969).

Let us take Kronecker's motto above as a 'minimal common denominator' of all-agreeable immutable truth(s). The laws of integer numbers are absolutely fixed and, as Thomas Aquinas (1225–74) asserted, not even God is capable of changing them. As Bertrand Russell noted, Peter Damian (1007–72), in the treatise 'On divine omnipotence', maintained that God can do things that are contrary to the law of contradiction, and can undo the past. This view was rejected by Aquinas and has since then been considered to be unorthodox (Russell 1989, p 407).

Furthermore, Kronecker's fundamental limitation is even more applicable to the physical Universe because of its fundamental and all-encompassing interconnectedness on all spatial and temporal scales. Yes, the fundamental physical constants (e.g. the mass ratios of elementary particles, like the above-mentioned $M/m$) are not guaranteed to remain fixed forever and may gradually change due to cosmological conditions. They can in no way acquire the same status of permanence as the ratio of, say, two given prime numbers of the digits of the pi number.

The mental 'locking' can go on in 'discrete portions' in, for example, the following way. Suppose we have 'matched' the $M/m$ ratio with a ratio of two large prime integers (it is always possible to do this with any degree of precision). Change in the 100th decimal digit 'induced by mental effort' will 'violate' this numerical locking and the $M/m$ ratio should now be 'better' approximated by the ratio of another pair of primes. This is an irreducible (though subtle) effect that Anderson wants to deny in any approximation. Note that for the philosophically oriented observer it is immaterial if the effect 'starts' at the 20th decimal digit or the 100th or 1000th digit. Due to chaotic (exponential) divergence (the butterfly effect), an arbitrary small effect can relatively quickly grow to an observable level. So, the apparently insignificant difference between very small effects (Jahn) and no effect at all (Anderson) turns into an acute dilemma of the 'either–or' category. Another opening for the above argument can be related to the introduction of the concepts of 'fuzzy' set theory. The unavoidable fuzziness of the space–time metric may place a fundamental limitation on precision (due to effects like relativistic metric fluctuations). Small 'intentional' perturbations may provide an additional component for this inherent fuzziness.

## 6.3 Premature and delayed discoveries

*Almost all new ideas have a certain aspect of foolishness when they are first produced.*

Alfred North Whitehead (1861–1947)

It is broadly accepted that many scientific discoveries, large and small, have been done 'ahead of their time'. The same is true for many scientific ideas. It is sufficient to recall Leonardo da Vinci's (1452–1519) predictions of air flight and submarines, or the mention of two small satellites of Mars by Jonathan Swift (1667–1745) in *Gulliver's Travels* (1726, amended in 1735). This was long before these satellites could be discovered by telescopes[1]. Likewise, some discoveries may be lost in the web of history, only to be rediscovered later (Teresi 2002) or—who knows?— perhaps not discovered at all if their time and the need for them has now gone.

However, there are many (perhaps, even more numerous) cases of the opposite kind, where discoveries and inventions were made much later than might haven been the case. One can call these 'delayed' or 'post-mature' discoveries. One can indicate three attributes of such delayed or post-mature discoveries.

(1) In retrospect, it must be judged to have been technically achievable at an earlier time using the methods available at that time.

---

[1] Swift's indications for their periods of revolution were quite close to what was discovered 150 years later (in 1877) for the two actual satellites of Mars—Phobos and Deimos.

(2) It must be judged to have been understandable and capable of being expressed in terms comprehensible to the educated public of that era.

(3) Its implications must have been capable of having been appreciated at the time.

Both pre-mature and delayed discoveries suggest a non-linear and complex model of the advancement of knowledge. However, pre-mature discoveries are either passively neglected or actively resisted at the time they are made. Discoveries can be pre-mature because they are conceptually misconnected with 'canonical knowledge', or are made by an obscure discoverer, or published in an obscure place (if at all), or are incompatible with the dominant religious and/or political doctrines of the time, etc.

It is a far trickier task to find a good explanation for delayed discoveries. The explanations here (if they are given at all) are usually not so obvious and are far less convincing. Some of the delayed discoveries have been postponed not by just a few years, but sometimes by many decades or even centuries. And here I am not talking about why the ancient Greeks did not invent television, or why Napoleon did not use his cell phone to call from Moscow to France. Instead, I want to mention a few such 'unperformed advancements' that were reasonably warranted by the conditions that existed long before they were actually discovered and/or announced. Others can probably add many more items to the following list of examples of 'late discoveries'.

## 6.4 The Möbius strip and recycling bins

*Aristotle maintained that women have fewer teeth than men; although he was twice married, it never occurred to him to verify this statement by examining his wives' mouths.*

Bertrand Russell

In my personal view one of the most striking examples of a delayed discovery is the famous Möbius strip (Möbius band). Ask anyone if is it is possible to draw a continuous line on both sides of a piece of paper without bending over the edge. Many people will, of course, answer affirmatively by referring to the famous Möbius strip. It is now widely known well beyond the mathematical community. There are numerous examples of its use in visual art; the imaginative drawings of the Dutch artist Maurits C Escher (1898–1972) are among the most popular of these.

So, this—by now well-known (almost anyone knows what it is)—construct of a one-sided surface was first described by the German mathematician August Ferdinand Möbius (1790–1868) in 1858 (as far as we know). Anyone can make it in a few minutes: just cut a rectangular paper strip of paper ABCD, bend it, connect it in a cross-over manner, B to D and A to C, and then glue or scotch tape it. It is a glorious Möbius strip: by contrast to the usual paper sheet it has only one surface (not two). Its topological properties are very interesting and quite different from the regular two-sided paper strip.

One can wonder whether perhaps it had been discovered before. Maybe it had. However, my (reasonably long) research on the subject failed to turn up a single piece of clear evidence that anyone prior to Möbius noticed the existence of a one-sided surface or used it in any artistic images (paintings, drawings, etc).

And this is despite the fact that every time we wrongly fasten our belts upside-down we in fact create an authentic replica of the Möbius strip. Thousands of people undoubtedly did this many times well before Möbius. And yet, it seemed to skip the attention of all the best minds from antiquity through the Renaissance and up to modern times. How could such an obvious thing, which can easily be made, understood and appreciated in a junior school, be so spectacularly unnoticed until some elderly German professor discovered it (most likely by mere chance) at a time when people were already building transcontinental railways and were on the verge of the commercial use of electricity?

Had not mankind produced Euclid, Leonardo, Pascal, Newton, Euler and many equally brilliant minds before? Why did none of them (at least to the best of our knowledge) devote a single line of their writings to mention such a simple and conceptually interesting construct as the Möbius strip certainly is?

To restate, the peculiar point which I am addressing here is the following: why was this simple construction first discovered in the middle of the nineteenth century (1858) and not earlier? Why did nobody mention it in any known writings before? Perhaps, even a curious child could 'discover' such a thing (just twist and fasten a belt the 'wrong way').

A similar question can be asked about many other things. For example, why were the steam engine, the telescope, the microscope, etc not invented in ancient Greece, Rome or Byzantium, or in dozens of other possible places? This is, indeed, strange. Really, why not? These things are, after all, so simple and obvious. In ancient Rome they already had children's toys that were set in motion by steam. So, why did nobody go one step further and construct a prototype of a steam engine? Why did this have to wait for another 15 centuries or so?

Likewise, the basic techniques needed for the construction of telescopes were also available for a long time. Glass polishing and the art of making corrective glasses (we call them 'spectacles' in English, 'lunettes' in French and 'ochki' in Russian) were known for centuries before somebody was smart enough to put convex and concave lenses together to discover the telescopic effect. Well, here we can still offer some (inadequate, perhaps) 'excuse' that it requires not just a certain technology (e.g. glass polishing) but also some theory to make a telescope and these two things might not easily be combined in the same person. But, again, the Möbius strip—no technology and no theory is needed to discover it.

Yes, this little Möbius strip does not require any technology—any paper, papyrus or even an elongated leaf will suffice. It could have been discovered in ancient Egypt, Babylon, Greece, Rome, China, India, Byzantium, Renaissance Europe, etc. But it was discovered only in the middle of the last century. How could such an obvious thing, which could easily be made, understood and appreciated by a child, remain unnoticed by everyone until the relatively obscure German mathematician discovered it in the middle of the 19th century? Why did none of the brilliant scientists and philosophers devote (at least to the best of our knowledge) a single line of their writings to mention such an obvious and topologically interesting thing as this odd Möbius strip?

For many viewers the Möbius strip produces an almost mesmerizing effect. This explains its frequent use in modern art. However, the popularity of the Möbius strip is not limited to art and mathematics. Nowadays it is often used as a universal symbol of recycling. Its circular (yet non-linear and 'twisted') shape evokes the process of transforming waste materials into useful resources. On a personal level, the Möbius strip represents a willingness to move with the constantly changing cycles in our life process, transforming our challenges into useful solutions. The Möbius strip also reveals planetary transformation. Historically, there have always been changes to the Earth, i.e., natural restructuring of land and ocean masses, continental drifts, tide changes, weather and climate changes, seasonal changes, etc. The Möbius strip shape is symbolic of the eternal change within the stillness itself.

On a more philosophical and esoteric level, the Möbius strip can be seen as an expression of non-duality (or, to put it better—the transcendence of duality). It reveals the unity of all polarities, creating a state of oneness, joining the whole and the part, the masculine and the feminine, expansion and contraction, spirit and matter, etc. Everything is one and nothing can be separated from anything else. All is completely intertwined, infinitely. The Möbius strip is a spiritually significant symbol of balance and union (yoga = union). The Buddhist philosophy of Tantrism is also expressed by the Möbius strip shape. 'Tantra' is continuity; the word is derived from the root 'tan', meaning to extend, extend continuously, to flow, to weave. The continuum is descriptive of the nature of reality, in the words of physicist David Bohm, 'a single unbroken wholeness in flowing movement' (Bohm 1980).

## 6.5 More delayed discoveries

The Möbius strip is just one of many examples of delayed discoveries. There are many equally amazing 'skips' in the history of science. There are quite remarkable examples even in the most basic of all the sciences, the arithmetic of integer numbers. Take, for example, prime numbers, of which there are many mentions in this book. Using the sieve of Eratosthenes (crossing out the multiples of 2,3,5,7,11, etc), one can easily find 'all' the prime numbers one by one. Theoretically, at least. Yes, the process is rather slow, but it appears almost self-evident that for integers between, say, one and 1000 all the primes should have been known since antiquity. Well—no.

As Leonard Dickson tells us in his *History of the Theory of Numbers* (Dickson 1952), medieval mathematicians believed that numbers in the form $[N = 2^{n-1}]$ are primes for every odd value of $n$ (here $2^n$ means two to the power of $n$). It is so easy to check that this is not so, and yet such a check was apparently not done until the late Renaissance period. And indeed, the above equation 'works' for $n = 3$, 5 and 7 (the values of $N$ are, respectively, 7, 31 and 127, all primes), but, as anyone can easily check (no calculator needed), it fails as early as $n = 9$ ($2^9 - 1 = 511 = 7 \times 73$).

It is then truly amazing that the latter (truly trivial) fact was first noticed by the mathematician Regius in 1536 (he also noticed that $2^{11} - 1 = 2047 = 23 \times 89$).

And here again, we are talking not about the dark ages, but of a time when Europe had already built its most magnificent cathedrals, and a time when book printing had already been a growth industry for almost a whole century. And to discover the above 'fact' (that $511 = 7 \times 73$) requires nothing but a piece of paper and less than an hour of test divisions (even the ancient Greeks knew how to do it).

It can be argued that one-sided surfaces and prime numbers are relatively abstract esoterica. Again, this depends on how one looks at it. But what about much more practical things? And here again, some amazing lapses can be found. Eyeglasses were well known in Europe from the 13th century on. Recall the churchman holding glasses painted by Jan Van Eyck in 1436. And yet, strange as it seems, it did not occur to anyone before 1605 to combine just two lenses (convex and concave) to produce the telescopic effect. Even if we accept, as some sources claim, that the principle of the telescope was known to Roger Bacon (1214–94), this is at best an example of an important instrument that was discovered only to be immediately forgotten. The whole of history might have turned out quite differently if the moons of Jupiter had been convincingly observed in the early Renaissance, say, in Dante's time (before the rise of the Inquisition), instead of some three centuries later via Galileo, who had the misfortune to come up with his telescope discoveries at the peak of the anti-humanist reaction.

Now take the microscope. This would probably be an even easier feat than the telescope. Just two convex lenses with a short focal distance (meaning strongly curved), this is all that is needed for a microscope. It would be rather primitive, of course, but it could still reveal many good (and bad) things like blood cells, bacteria, etc. The quality of glass polishing needed for such lenses was around for centuries before the Dutchman Zaccharias Janssen and his son Hans built the first microscope by placing two lenses in a tube. That was in 1590, three centuries after the Italian Salvino D'Armate made the first wearable eyeglasses (around 1284).

The steam engine (not just as a mere toy), man-carrying hot air balloons (the Montgolfier brothers, 1783), a quantitative scale for temperature, and atmospheric pressure—all these things could quite naturally be expected to have been discovered or constructed centuries earlier than they were actually delivered. For example, it would have been technically possible to build (and broadly use) hot air balloons for human flight in any major civilization from Pharaonic Egypt on. On the more theoretic side, the law of the pendulum (the fact that the squared period is a linear function of the pendulum's length) could easily have been established in the ancient period. All it would take would be two people simultaneously counting how many times two bobs hanging on ropes of different lengths (one length is fixed, the second varies) swing during a given time interval. After several repeats simple data plotting could have revealed the pendulum law thousands of years before Christian Huygens (1629–95) discovered it.

But of course, this all happened long ago. Now we are not going to miss anything significant. But are we? Is the fear of simplicity an inherent part of our mentality and culture? Could it be one of our other hidden curses?

Let us imagine that by some miracle the Möbius strip had not been discovered until today. Who would dare to dream that in the current peer review system a grant proposal to 'search for one-sided connected surfaces in a three-dimensional space' would have any chance of passing and gaining approval for research funding? The Möbius strip is too simple for peer review to digest. Most likely the experts would right away dismiss the idea as impossible rubbish.

These few random examples from whatever time tell us that it is often the obvious thing that is the most difficult thing to notice and appreciate as something special. This is probably the main explanation for the existence of numerous delayed discoveries. And some of them may likely still be on their way to us.

To wrap up this discourse, it should be mentioned (again and again) that it is a rather general and deeply rooted phenomenon that a substantial segment of the scientific establishment is at best utterly unreceptive and at worst openly hostile to a profoundly new idea or approach. It happened not in the dark ages, but around 1977, that Mitchell Feigenbaum could not publish his epochal ideas on the universality of chaos for a few years because of repeated rejections by several major physical journals. Likewise, it was only after some ten rejections that I published the idea of isotopic fiber optics as an alternative to conventional fiber optics (Berezin 1989). Perhaps the still-existing lack of genuine appreciation for stable isotopic diversity and the technological potential of isotopicity falls into the same category of ideas that were spelled out before their 'right' time had come. But again, who here can determine what is the 'right time' for an idea or a discovery? Only the future can tell.

The term 'disruptive technology' is often used in many contexts. It seems that nowadays nanotechnology is most often labeled with this term. Like any new disruptive technology, nanotechnology has its redeeming as well as its potentially hazardous aspects. While the benefits may appear more obvious than the pitfalls, both need to be properly appreciated. Likewise, isotopic nanotechnology (if such a term becomes accepted) could face the same challenges as standard nanotechnology, but also a few isotopically specific ones. As for common nanotechnology, aspects of potential concern may range from fears of pervasively intrusive Orwellian surveillance nano-devices through to almost apocalyptic scenarios of 'grey goo' molecular assemblers going out of human control and devouring everything around them.

## 6.6 Big science and peer review

*Faced with the choice between changing one's mind and proving that there is no need to do so, almost everyone gets busy on the proof.*

John Kenneth Galbraith (1908–2006)

The modern research funding system is generally oppressive to creativity and innovation. More specifically, expert peer review favors fashionable projects with predictable outcomes and usually rejects novel higher risk proposals; the system mostly supports 'me too' projects ('do as others do') and encourages mediocrity and triviality rather than true innovation, and the latter gets only lip service. It encourages the proliferation of powerful controlling cliques that operate as old boys' clubs ('I'll fund you, if you fund me'); it fosters quasi-feudal research empires, with 'soft money' researchers (mostly postdocs) doing all the real work while tenured professors are engaged largely on the conference circuit and grantsmanship; research supervisors often lose touch with the experimental bench, and yet remain in full control of their laboratories' budgets and operations; and such an imperial model of operation discourages the creativity of junior researchers and their ideas are often misappropriated by supervisors. These factors generally render the modern research funding system financially wasteful and resentful of public accountability.

The core defect of the American/Canadian research funding system is that it is based on the idea of competition between proposals. The proposals are scored by 'experts' (anonymous peer reviewers) and the scores are used to separate them into funded and unfunded piles. This is how the system operates, at least in its own description of itself.

Although the policy is superficially justified by peer review quality control, in practice it almost invariably favors research along well-established lines and discourages novel approaches, real innovation and risk taking. This state of affairs encourages researchers to spend much time and energy on writing grant proposals, so-called grantsmanship (with energy taken away, of course, from research). This leads to the proliferation of conservatism, cronyism and the over-concentration of research funds in the hands of elite and control groups. Instead of being a genuine (interactive) competition, it is, in fact, a centralized and authoritarian grant distribution system. The net result is the marginalization of truly exploratory research that often has a significant level of uncertainty and a noticeably higher failure rate than sure-fire science.

In 1993, Noble Prize physicist Leon Lederman had already noticed (Lederman 1993) that young scientists spend up to 40% of their time seeking research funding and writing grant proposals. It is likely that the figure is worse now. It is time to stop this silliness. In terms of time stolen from productive research this irresponsible waste of human talent to write and read uncountable truckloads of 'proposals' amounts annually to a colossal waste of money and time. It is helpful to understand why (contrary to its superficial 'obviousness') the idea of peer review of proposals is a red herring (Osmond (1983), Gordon (1993), Forsdyke (1993), (2000)), especially when it is entertained in an environment of fierce selectivity where most proposals end up in a dustbin.

We live in what could be called a 'culture of presumed incompetence'. At the starting line the default assumption is that grant applicants are incompetent *unless* they manage to prove otherwise (in the eyes of 'experts', of course). Furthermore, they are required to keep busy with proving this over and over again each time they reapply for a new grant. This practice is a sharp departure from most other specialized areas in which, after a period of proper training, apprenticeship and (when required) certification, the practitioner is deemed to be capable of handling his/her trade at least fairly well.

Nobody requests that a surgeon should justify on paper his/her intentions before he/she is allowed to enter the operating theater. His/her ability to act professionally and independently is trusted on the basis of prior training and a sustained record of practice. Yet, such trust is precisely what is denied to researchers when they are regularly forced to submit applications with a detailed description of yet-to-be done work. How far would Newton or Darwin have gotten if they had to justify their ideas before they were allowed to work on them?

For all practical purposes it is fully sufficient to have a one-page proposal (written largely for filing purposes) amended with personal data assessed by a simple screening committee many times smaller than the present grant councils. The assessment (to determine the funding level) should largely concentrate on the track record that shows the *actual* accomplishments of the researcher. The assessment of *already performed* work is undoubtedly a much more reliable activity than the futurological crystal-balling of proposals (promises).

Nowhere (except perhaps for some very specific product development) do detailed proposals make any real sense. Quoting Nobel Laureate Szent-Gyorgyi: 'Writing proposals was always an agony to me. I always tried to live up to the commandment, "don't lie if you don't have to." I had to [...] A discovery must be, by definition, at variance with existing knowledge. During my lifetime, I made two. Both were rejected out of hand by the popes of the field' (Szent-Gyorgyi 1972). In the words of biophysicist Richard Gordon (Gordon 1993), 'We are forced to lie to obtain funds to seek truth'.

Fortunately (and justifiably) there is quite strong criticism of peer review inside the research community (e.g. Osmond (1983), Kenward (1984), Forsdyke (1993), (2000), Horrobin (1990), (1996), Savan (1988), Foltz (2000)). On the other hand, it is instructive to look at the arguments of the proponents of the heavy-duty peer review selectivity system. Ronald Kostoff lists such factors as the 'commitment to high-quality reviews' and 'motivation to conduct a technically credible review' on the 'highest standards in selecting reviewers' and 'the reviewers' competence and objectivity' (Kostoff 1997).

The problem is that all these terms sound great but in essence they are too nebulous, if not fantastic. The highest standards, credibility, objectivity, etc—these are subjective and circularly defined terms which are of little real use without an interpretive and authoritative body implementing them. And this returns us to square one, with the net result that the 'encouragement of corruption is an inevitable consequence of review by expert competitors' (Horrobin 1996).

## 6.7 Orthodoxies and heresies

*Metaphysics means nothing but an unusually obstinate effort to think clearly.*
William James (1842–1910)

Many topics in off-mainstream scientific directions have been referred to as examples of 'pseudo-science', 'pathological science', 'metaphysics', or 'paranormal science' (e.g. Rousseau 1992, Berezin 1996). These terms have not proved to be particularly helpful in promoting a rational and constructive dialogue between the various parties involved in the discussion. Yet some suggest that it might be both more realistic and more constructive to regard these controversial topics as scientific 'heresies'. After all, a heresy generally can be understood as being a proposition, directed at a profession or other organization, that is both a challenge to (currently consensual) understanding and a challenge to power. A heresy, therefore, has both intellectual and political content.

Some of the (more-or-less recent) scientific heresies have connections to 'eternal' problems, or at least, to some aspects of them. Some examples of these are given below.

*The red shift controversy*—is the red shift in the spectral lines of distant galaxies indeed an indication that all the Universe (at least, the observable, or 'Big Bang Universe, BBU) originated from a cosmic singularity (Big Bang) some 10–15 billions years ago? Or can we put on the table some alternatives to the BBU explanations for consideration? For example, could (what we perceive as) the red shift be due to some effect(s) of 'photon aging' and/or differences of the speed of light at various parts of the Universe, or, perhaps, some other possible effects or 'mechanisms'? Or, could the red shift be interpreted within the framework of a steady state cosmology according to which the Universe remains more-or-less 'the same' for an indefinite time (Fred Hoyle)? Likewise, it could well be the case that some other hypotheses regarding the origin and nature of the red shift are presently under consideration that I am not aware of. Nobody can know everything, so I do not need to make any apologies for something that may well be outside my scope or vision.

*The cold fusion controversy*—this exploded in 1989 with the announcement by two distinguished electrochemists (Stanley Pons and Martin Fleischmann) that they had observed a chemically stimulated nuclear reaction (some claims to that effect had been known in earlier times, but they did not produce a significant impact on the research community). The societal impact of this seemingly purely 'scientific' controversy is that it challenges the notion of the stability of matter and (if confirmed) gives some validation to the claims of alchemy and other acts commonly labeled as magic. I talked more about that in section 4.16 of this book. Naturally, the scientific community (for the most part) remains highly skeptical, despite the fact that on purely scientific grounds the issue remains open at the time of writing (2016) and new claims to that effect keep appearing in the media.

*A variety of parapsychological claims*—most of these appeal to this-or-that form of mind–matter interactions, which have no commonly accepted explanation within the mainstream framework. They include, for instance, the experimental studies by Robert Jahn's Engineering Anomalies Research Laboratory at Princeton University (Jahn and Dunne (1988), see also Anderson (1990), (1991)) and such ideas as long-distance interactions between animals through 'morphic resonance' (Sheldrake (1988), (2012), McTaggart (2003)).

*UFO claims, crop circles, etc*—numerous claims of UFO (unidentified flying objects) sightings have been neither refuted nor solidly confirmed. The available interpretations range from extra-terrestrials and ghost-related phenomena to a flat denial and/or explanations based on some (often not completely understood) mundane effects, such as optical illusions or electrical discharges like ball lightning, etc. Links between these phenomena and 'eternal' problems are sometime claimed, but, for the most part, they remain highly speculative.

As the author of this book, I do not voice any definite opinion here about the validity of any of the above issues (and many more can be added to the above list). People, generally, have a broad variety of opinions in this regard. My own attitude to many of these claims was (and still is) of a 'maybe' or 'why not?' type of questioning rather than automatic dismissal. As has been mentioned in this book more than once, in the Universe we are in, only the laws of mathematics cannot be changed (e.g. making $2 + 2 = 5$ or changing the list of prime numbers). As for the 'physical' Universe, every possibility remains open. Perhaps, even the law of energy conservation can be circumvented, for example, by extracting energy (as some suggest) from the expansion of the Universe—and yes, who knows, maybe indeed some day we will learn how to do it.

There is not, of course, nor likely can there be, any universally valid 'mechanism' for demarcation of the boundaries between mainstream and fringe ideas. For the lack of better tools, the commonly adopted mechanism for this is peer review in science. It is usually justified by arguments that are akin to Winston Churchill's famous observation about democracy, that it is 'the worst possible system, except for all the others'. Nonetheless, in many cases peer review does not live up even to this Churchillian definition and on the ground it acts (to a large extent, at least) as a mechanism of ideological suppression and censorship. Despite the fact that issues like peer review, peer pressure, etc, seem to be of peripheral importance for the 'eternal' and 'metaphysical' questions, they directly and indirectly affect the path of thinking of many people, as was discussed by the iconoclastic philosopher Paul Feyerabend (1924–94) in his provocatively titled essay 'How to defend society against science' (Feyerabend 1975).

An example of how a mainstream topic can be conveniently turned into a fringe one can be provided by paraphrasing the parable of 'are automobiles alive?' (Tipler 1994). The (more-or-less mainstream conceptual) definition of life is 'a system is alive if it interacts with the environment, is capable of reproducing itself and is preserved by natural selection'. Automobiles do precisely this: they reproduce in factories using human mechanics as their 'environmental arrangement'. The form of automobiles in their environment is preserved by natural selection: there is a fierce struggle for existence between various brands of automobiles and various car manufacturers. By this definition of life, not only are automobiles alive, but all machines—in particular computers—are (Tipler 1994). Furthermore, according to Roger Penrose (Penrose 1994), the phenomenon of (human) consciousness may be fundamentally related to (sub-) quantum physics. Penrose proposes that consciousness can be fundamentally non-computable, i.e. cannot be adequately modeled on any (digital) computer.

And if we really look without any pre-established bias at the *real* record of modern mainstream and established science regarding what it *really* can tell us about the most fundamental questions, the record does not look too encouraging. This is how the biologist Robert Lanza summarized the situation in his bestseller, *Biocentrism*.

Classic science answers to basic questions:

How did the Big Bang happen?—*Unknown.*
What was the Big Bang?—*Unknown.*

What, if anything, existed before the Big Bang?—*Unknown*.
What is the nature of dark energy, the dominant entity of the cosmos?—*Unknown*.
What is the nature of dark matter, the second most prevalent entity?—*Unknown*.
How did life arise?—*Unknown*.
How did consciousness arise?—*Unknown*.
What is the nature of consciousness?—*Unknown*.
What is the fate of the universe; for example, will it keep expanding?—*Seemingly yes*.
Why are the constants the way they are?—*Unknown*.
Why are there exactly four forces?—*Unknown*.
Is life further experienced after one's body dies?—*Unknown*.
Which book provides the best answers?—*There is no single book*.

Okay, so what can science tell us? A lot—there are libraries full of knowledge. All of it has to do with classifications and sub-classifications of all manner of objects, living and non-living, and categorizations of their properties, such as the ductility and strength of steel versus copper, and how processes work, such as how stars are born and how viruses replicate. In short, science seeks to discover the properties and processes within the cosmos. How to form metals into bridges, how to build an airplane, how to perform reconstructive surgery—science is peerless at the things we need to make everyday life easier. So those who ask science to provide the ultimate answers or to explain the fundamentals of existence are looking in the wrong place—it is like asking particle physics to evaluate art. Scientists do not admit to this, however. Branches of science such as cosmology act as if science can indeed provide answers in the deepest bedrock areas of inquiry, and its success in the established pantheon of other endeavors has let all of us say, 'Go ahead, give it a go.' But thus far, it has had little or no success.
(Lanza and Berman 2009, pp 155–6)

Yes, the record of science concerning the fundamental questions of existence indeed seems quite miserable. And yet, in no way should the above 'devastating' quotation be taken as a discouragement of our continuous quests on these matters. However, the angle of these quests should not, perhaps, be confined to the mental straitjackets of the orthodoxies of mainstream science. To that end, recent (and often fascinating) speculations regarding the 'world as simulated reality' and such ideas as the 'matrix' and artificial (machine-based) consciousness have stirred a high-level philosophical discourse (Bostrom (2003), Chalmers (2005), Kurzweil (1999), (2005), Minsky (1994), Moravec (1999), Vinge (1993), Crawford (2000)), which sometimes projects itself into metaphysical and eschatological realms (e.g. Leslie 1996, Lewis 1986, Tipler 1989, Webb 2002, Rees 2003, Fukuyama 2004), not to mention the associated science-fiction literature (e.g. Egan 1994, 1997) and even some earlier writings of now-almost-forgotten authors (e.g. Feodorov 1913, Condorcet 1955 (original 1795)).

## 6.8 The crowd mentality and the interdisciplinary paradox

*If you want to make enemies, try to change something.*
Woodrow Wilson (1856–1924),
28th president of the United States (1913–1921)

In spite of the great complexity of the modern research system, its operation can to a large extent be illustrated by the following short scenario formulated as a self-testing question. Imagine an unknown young physicist without an institutional affiliation or a professional track record who submits some papers proposing a radically new view of the physical Universe to key physical journals and also applies to an established research funding agency for a modest grant to facilitate further research in this direction. Ask yourself the following pair of questions. What chance is there that a major physical journal will accept a paper from an unknown researcher? And what chance is there of this physicist being awarded even a very small (seed) research grant? Both questions were, in fact, posed to several groups of academics, with an invariably predictable result. Essentially, all the respondents ranked the chances of the said young physicist as close to zero. Only by gradually working through the system and playing by the rules (which normally takes many years) would this aspiring physicist attain a reasonable chance of scoring on both counts.

Needless to say, the above scenario almost exactly fits the description of the young Albert Einstein around the year 1905, when he was about to complete his first presentation of the special relativity theory. A hundred years later the only noticeable corrective to the picture would probably be the possibility of instant publication of his/her writings on the internet and/or via self-publishing. However, easy as they may be to arrange, internet posting and self-publishing have virtually no bearing on the acceptance of articles by a 'real' (reputable) journal or obtaining research funding.

The above clash between innovative ideas and the authority of established experts can be corroborated by many other examples. Most people are familiar with the numerous mind-boggling blunders associated with predictions made by experts, such as the statement that 'no stones can ever fall from the sky' quoted in section 5.6. After a fall of stones (meteorites) in France on 24 July 1790, more than 300 written statements were sent to the French Academy, together with pieces of the stones, but the academicians merely ridiculed the ignorance of the writers and dismissed what they called a 'physically impossible phenomenon' (King-Hele 1975).

As with most such broad (umbrella) concepts, 'innovation' is difficult to define in unambiguously inclusive/exclusive terms. But in the context of university (academic) research, a typical working definition of innovation is generally based on a small number of criteria. As a rule, the acceptance of research as innovative by a research (sub-) community requires that it meet at least some of the following requirements: it should attract attention and interest from a sufficiently large proportion of the (sub-) community; it should be performed and/or directed by people who already have authority in the field, and who are generally recognized as established experts; and it should have a significant number of impressive 'visible' (usually easily quantifiable) parameters, such as numbers of associated journal and conference publications, a high level of research funding, several graduate students and postdoctoral associates, etc. In short, in order to succeed you already have to be successful. It sounds pretty much like the well-known saying that if you want to be rich all you need is to have a lot of money.

This principle is often cast in the form of the 'St Matthew effect' (as the Bible puts it, 'those who have get more and those without will not get anything'), as quoted by many authors (Savan (1988), Horrobin (1990), (1996), Cole (1992), Forsdyke (1993), (2000), Foltz (2000)). Almost anyone in the research community would be likely to agree that the above criteria are not without some pragmatic worth. After all, there are numerous examples of truly groundbreaking research indeed satisfying the above conditions. Equally important is that they provide some structural stability to the system. To do away with them suddenly and completely might appear to be overly disruptive. Yet at the same time it is equally likely that most people who have ever had their hands on research will be able to point to the severe limitations of these criteria, and they will often be able to come up with striking counterexamples. Thus potentially pivotal research may simply be too far ahead of its time and the accepted baseline of 'common sense' in a particular field. It can then be either stonewalled, like Alfred Wegener's theory of continental drift (Oreskes 1999), or simply face a complete lack of interest (who was interested in Gregor Mendel's genetic experiments at the time they were performed?).

That a tough ride is given to new ideas has, of course, been a common trait for much of human history. For better or worse, people are almost without exception suspicious of novelty and originality. Despite the fact that a reasonable dose of skepticism is often justified, the present organizational and managerial structure of the research system is, in general, oppressive towards innovative ideas and is therefore in need of some major changes (Berezin 2001).

Much of the novelty and originality comes from research that is at the interfaces of a variety of disciplines and hence can often be seen as 'interdisciplinary'. Although superficially interdisciplinary research is welcome in academia and much kudos is given to it, in reality it more often than not falls into the grey area of suspicion. As an interdisciplinary scientist myself, more than once I was told 'Alex, you have too many papers on too many topics', and I was asked 'Can you not concentrate on one particular topic?' However, to be fair, a genuinely positive appreciation of interdisciplinary work is not rare inside or outside academia.

## 6.9 Going 'around' the system: isotopicity as an example

Through the course of history, innovation has always been a mixture of several key factors, such as curiosity and ingenuity, individual creativity, and the promise of personal reward. The relative importance of these factors for different individuals and different circumstances of course varies widely. Such commonly held innovation incentives as copyright and patents are facing criticism (Mann 1998, Shulman 2000). There are growing concerns that these instruments are increasingly becoming roadblocks on the path of innovation. In the final analysis it is curiosity, creativity and ingenuity that are of overridingly great importance in producing innovation, rather than actual or anticipated rewards. The whole history of science confirms this. Yet these are often delicately balanced individual qualities that can easily be knocked out, or even killed, by insensitivity, indifference and scorn.

One might wonder why isotopic engineering did not evolve to the level and scale of a generally recognized and socially visible technological activity.

On the economic side, one of the reasons for the current low level of research activity in this area might be the cost of pure stable isotopes. The existing methods for isotope separation (like molecular beam technology) make pure isotopes quite expensive, especially for academy-based researchers whose funding mostly comes from research grants. Unlike the isotope separation technology for nuclear applications, whether for peaceful (nuclear power and medicine) or military (nuclear weapons) use, the stable isotope separation industry is relatively limited. Should major applications of stable isotopes for micro- and nano-technology became a reality, we are likely to see a drop in the costs for stable isotopes due to the economic mechanisms and market laws of mass production.

Likewise, on the technical side, the potential benefits of isotopic technology— which within the context of this book mainly refer to the use of the purposeful structuring of stable isotopes—are still likely waiting to be unfolded. It is sufficient to mention here that most of the nano-technological gadgets that relate to biology and medicine, such as molecular assemblers or nanobots (nano-robots), could well incorporate atomic-scale isotopic information storage as the most compact and informationally dense means of carrying algorithms for the execution of instructions. The latter may actually be miniature isotopically based quantum computers monitoring such processes as self-assembly and replication. Meanwhile, as was mentioned in section 4.7, some progress has been made on isotopic engineering (e.g. Berezin 1989, Plekhanov 2004, Haller 1995), but work in this area remains somewhat sporadic and still awaits its full unfolding, especially on the biological and biomedical frontiers.

## 6.10 Random creativity and Laputa machines

*Civilization advances by extending the number of important operations which we can perform without thinking of them.*

Alfred North Whitehead (1861–1947)

In *Gulliver's Travels*, Jonathan Swift colorfully describes Gulliver's visit to the flying island of Laputa. It is populated by some highly eccentric people, of whom many are scientists. How do they produce their scientific results? They have machines with many rotating cubes on which various scientific terms are written. By randomly rotating these cubes various scientific terms come up, and that is how new scientific ideas are created—by the random combination of terms. This is a kind 'scientific analogue' of tarot card spreads and there are certainly many possible versions of such a technology.

Now, in our computer age, we may be traveling a few steps further. There are now programs in existence (they can be found easily on the web) for the creation of scientific (or rather scientific-looking) papers. There is very little doubt that with the growing sophistication of computer systems these computer-generated papers (and 'ideas') will be practically indistinguishable from the 'real' ones (perhaps they already are).

However, this may not be an entirely negative development. The use of randomness in many areas of human life and activity has gone on for centuries and not without some examples of success. Doges (chief magistrates) in Venice were appointed at random from the pool of eligible candidates, and Venice did not do particularly badly in the course of history. Imagine the American presidential election being conducted as a random draw from among a few finalists. Political campaigning would probably be much less hostile, and healthier, more constructive and cooperative, and much less loaded with false promises and mutual bickering. Of course, proper safeguards would be needed to assure genuine randomness and prevent tampering or cheating, but that, in principle, could be done (perhaps using the isotopic random number generators discussed in section 4.14). The role of randomized stimulation was recently emphasized by Edward De Bono, a writer on human creativity (de Bono 1985). One may also recall an earlier bestseller, *The Dice Man* by Luke Rhinehart, where the protagonist made all his decisions by tossing dice (Rhinehart 1971). Much art is likely and/or seemingly produced by random combinations of elements (e.g. Dali and other surrealists, the imaginary arts, etc).

Likewise, a random grant system would be enormously cheaper to run than the present funding bureaucracy. And in terms of the results it would probably be better, not worse, than the present highly selective (and hence equally arbitrary) funding system. At least a random model is free from the conformist pressures characteristic of any peer review system. Innovative research has a better chance in a random system than under the pressures of a peer review mechanism. To avoid the extremes, the applicants, of course, should satisfy criteria based on their credentials and track record (there is no need for an in-depth peer review to establish this) and random awards should vary in amounts in such a way as to avoid zero funding for active research programs (Berezin 2001).

## 6.11 Living in the Matrix—physics reloaded

*There are known knowns. These are things we know that we know.*
*There are known unknowns. That is to say, there are things that we know we*
*don't know. But there are also unknown unknowns.*
*There are things we don't know we don't know.*

　　　　　　　Donald Rumsfeld (b 1932), former US Secretary of Defense

In line with the 'digital information' tone of this book, I would like to inject a spirit of radical thinking and futurology here by discussing the so-called 'simulation argument'. While it lies at the boundary of science fiction and real computer science, it is gaining some momentum along the lines of the transhumanism and posthumanism discourse (Berezin 2006).

The rapid evolution of computers and the exponential growth of computing power (Moore's law) have triggered some interesting philosophical speculations (that actually can be traced back to even earlier times). Recently, the philosopher and futurologist Nick Bostrom in an article with the provocative title 'Are you living in a computer simulation?' (Bostrom 2003) presented a modern version of the argument for solipsism and subjective idealism (see, e.g. George Berkeley for earlier versions), according to which the outside world is indistinguishable from illusion. This simulation argument resonates with several other recent developments, such as the discourse on whether future computers will be able to emulate human consciousness (artificial intelligence), whether future humans will reach a biological symbiosis with computers (the stage of transhumanism or posthumanism, as defined below), or (in a more radical prediction) whether humans will actually be replaced by artificial self-replicating life forms (e.g. 'silicon life').

These philosophical themes have even attained some notorious visibility in mass culture and entertainment (e.g. the recent blockbuster movies *Matrix* and *Matrix Reloaded*). An earlier reflection on these issues was presented by Frank Tipler in his 1994 book, *The Physics of Immortality* (Tipler, 1994). His argument of the 'omega point' was akin to the earlier ideas of universal convergence that were developed by the French Jesuit and paleontologist Pierre Teilhard de Chardin (1881–1955). Although the latter did not talk explicitly about the simulation hypothesis, the extrapolation of his ideas potentially points in this direction.

Even before Bostrom's articles, simulation ideas were well represented in the science fiction literature. Without going into a detailed discussion of much of this work (it could easily take many pages), I will mention here the two dystopian novels by Greg Egan: *Permutation City* (Egan 1994) and *Diaspora* (Egan 1997). They address anticipated events in the middle of the 21st century. Computing power has increased enormously, to the extent that it is possible to run very detailed simulations of human brains and human bodies. Also, scanning technology has improved to the point where it is possible to scan existing brains at the atomic level. In *Permutation City* Egan describes a community of simulated humans (Copies) with multiple copies of the 'same' individuals. But it is not all good news: many Copies cannot tolerate this state of affairs and attempt to 'bail out' into 'reality'. The Copies who do continue as simulations are worried about their rights and about maintaining access to the computing power on which their very 'existence' depends.

In *Diaspora* humanity is divided into three types: 'polis citizens' (downloaded minds running on software in virtual reality), 'Gleisner robots' (software people living embodied lives in robot bodies) and 'fleshers', people still embodied in good old-fashioned meat. The novels deal with many existential, philosophical and mathematical issues that would likely arise in the realm of grand computer simulations (like, for example, what would we experience if our brains were computer-upgraded to perceive a 12-dimensional world?).

Partly because of the success of the *Matrix* movie series, many aspects of the simulation argument have been discussed recently from a variety of philosophical positions. David Chalmers in his work *The Matrix as Metaphysics* (Chalmers 2005) posited that the matrix presents a version of an old-philosophical fable: the brain in a vat. The brain is stimulated with the same sort of inputs that a normal embodied brain receives. To do this, the brain is connected to a giant computer simulation of a world. The simulation determines which inputs the brain receives. When the brain produces outputs, these are fed back into the simulation. The internal state of the brain is just like that of a normal brain, despite the fact that it lacks a body. From the brain's point of view, things seem very much as they seem to all of us. The brain is massively deluded, it seems. It has all sorts of false beliefs about the world. It believes that it has a body, but it has no body. It believes that it is walking outside in the sunlight, but in fact it is inside a dark laboratory.

Furthermore, even the existence of the 'real' (biological) brain is not essential—the entire simulation can proceed as a chain of bits (0s and 1s) in some super-computer. As Nick Bostrom suggested (Bostrom 2003), it is not out of the question that in the history of the Universe technology will evolve that will allow beings to create computer simulations of entire worlds. There may well be vast numbers of such computer simulations, compared to just one real world. If so, there may well be many more beings who are in a matrix than beings who are not. Given all of this, one might even infer that it is more likely that we are in a matrix than that we are not. Whether this is right or wrong, it certainly seems that we cannot be certain that we are not in a matrix.

As Chalmers points out (Chalmers 2005), the question of 'reality' is a tricky one. We should not necessarily see a simulated world as being 'unreal'—it just exists on the basis of a different metaphysical substrate, computer 'bits' instead of atoms and molecules. Therefore, according to Chalmers, the situation (even if we live in the matrix) may not be that worrisome—the simulated world just has a more fundamental metaphysical substrate (digital bits and information) than ordinary physical matter (electrons, atoms and molecules). In short, whereas the 'ordinary' physical world is based on the dynamics of material particles, the simulated world has its foundation in an unstoppable play of computer digits. In a metaphorical way one could say that such a simulated world would be a direct projection of the ideal Platonic world of numbers.

It is interesting to note that notwithstanding the speculativeness of the simulation argument, we do have an obvious (and 'real') analogy to it in our everyday life. Most (all?) people have dreams during their sleep and sometimes these dreams are very vivid. Assuming that we can envision our brain as a (biological) computer, can we say that it actually creates a simulated reality during our dreams? In our dreams do 'we' remain our 'real selves', or do we emanate our simulated twins? Amazingly, analysis of the nature of our dreams leads to almost the same questions as are raised by the simulation argument.

Physicist Paul Davis puts it this way: 'We are fascinated by dreams. Those people who, like myself, dream very vividly often have the experience of being 'trapped' in a dream that we believe is real [...] Can we be absolutely sure that the "dream world" is illusory and the "awake world" real? Could it be the other way about, or that both are real, or neither?' (Davis 1993, p 117).

Similar dream-or-reality ideas are plentiful in science fiction. While not attempting to review this truly impressive body of literature, I will mention here the novel *Solaris* (1961) by the renowned Polish writer Stanislaw Lem (English translation: Lem 1981), which has been adapted for the screen several times (the best known movie was made by the Russian director Andrei Tarkovsky). The novel talks about the discovery of a gigantic sentient colloidal ocean on the planet Solaris. The planet-scale brain is capable of incredible self-regulation, governing its macro-processes by controlling its orbit around two suns, and also its micro-processes by the manipulation of neutrino fields to create phantasmic simulacra of human beings.

When the novel's protagonist, psychologist Kris Kelvin, arrives at the Solaris research station orbiting the planet, he, to his amazement and horror, encounters a visitor—Rheya, a simulacrum of his dead wife, for whose suicide back on Earth he has blamed himself for several years. The scientists on the Solaris station come to realize that the sentient ocean is capable of producing materialized forms of the scientists' own unconscious thoughts. Thus, the simulated Rheya is actually a materialized projection of what Kelvin's memory contained about the actual Rheya back on Earth. But physically the new Rheya is different: as Kelvin discovers through the atomic analysis of her blood, she does not consist of regular atoms but is actually made up of a system of stabilized neutrinos organized by a quantum field which is apparently under the planet's control. However, her behavioral autonomy is isomorphic with human behavior.

Thus, neutrino beings created by the ocean Solaris can be seen as intermediate between 'real' people (made of atoms and molecules) and 'true' computer simulations which 'exist' in the form of electronic bits in a running computer program. And this brings us to the following question: what grounds (if any) do we have to say that beings made of atoms are 'real' while beings in the form of electronic bits are 'fictional'? While not pretending to give a definite answer to this question (perhaps, there is not one), I believe that this opens up a new and fascinating avenue to ponder on the ultimate issues of human and universal existence.

There are many other examples of items of mass entertainment that deal with the interface between what is 'real' and what is 'simulated'. Such popular television series as *The Outer Limits* or *The Twilight Zone* and the impressive success of such books as the *Harry Potter* series and *The Lord of the Rings* trilogy evidence the actuality and importance of the theme of 'reality versus dream world' for people in general. In fact, many of these fictional books and movies address questions of ultimate reality and meaning directly. Some of the typical questions are given below.

– Does it really matter if we are simulated reality or not?
– Do simulated beings (if there are any) belong to the same moral Universe as us, presumably 'real', beings?
– To what extent can our dreams and imaginary worlds be considered to be simulated realities?
– Do the quality aspects of simulation affect the ethical values we assign to them?
– Is it morally prohibited to simulate 'anti-reality', meaning by that some ultimately evil Universes of discourse? (It is almost certain that if a 'positive' reality can be simulated, a 'negative' one can also be simulated).

- If a strong ethical component in the simulated acts can be indicated, what are the criteria or guiding principles there?
- By whose moral authority are these presumed criteria and principles to be imposed?

By their very nature, most of these questions are open-ended. It is very difficult, if not impossible, to provide an unambiguous single answer to most of them, unless we are willing to confine ourselves to a stringent and limited philosophical frame. At the same time, there are some areas where the interface between the real and the simulated can be (to a degree, of course) explored at the experiential level. Numerous observations (and even some in-depth studies) on drug-induced hallucinations, enlightenment and meditation practices, and Eastern and Western mysticism and spiritual awareness, although diverse in quality, intentions and results—all point to the expectation that our understanding of the interface between the real and the imaginary (or simulated) is likely to attain new levels. Perhaps a useful connection to explore in the context of these questions can be provided by the notion of the co-existence of parallel Universes and multiple realities—ideas that are currently gaining strength at the advanced frontier of quantum physics and cosmology (Berezin 2004, Randall 2005).

Or, in other words, is it, even in principle, possible to tell whether we are in a simulated reality or a real one? Is there any difference between the two? Does it matter? And how should we behave if we know that we were living in a simulated reality?

Recently the idea that 'the Universe is a huge computer' has become popular in mainstream physics (Lloyd 2006). This view entails that all physical processes (atomic dynamics, biology, cosmological events, etc) are elements of a single computational process. In this vein, the simulation argument (whether it uses the image of the matrix or another relevant metaphor) would become an organic part of the underlying metaphysics rather than an odd addition to it. In a theological interpretation, the same, in essence, can be stated, as if physical reality is represented in the mind of God, and our own thoughts and perceptions depend on God's mind. In such a view the simulation of the world is implemented in the mind of God. If this is right, we should say that physical processes really exist: it is just that at the most fundamental level they are constituted by processes in the mind of God.

In splitting his arguments into several alternative scenarios, Bostrom posits that with a very high likelihood at least one of the following three propositions is true: (1) the human species is very likely to become extinct before reaching the post-human stage; (2) any post-human civilization is extremely unlikely to run a significant number of simulations of their evolutionary history (or variations thereof); and (3) we are almost certainly living in a computer simulation.

The first of these propositions is another form of 'doomsday argument', according to which there is a very high probability that human civilization (and humanity itself) will be destroyed within a historically short time frame. While the doomsday argument has an implicit (sometimes explicit) presence in most key religions (say, the Book of Revelation in the Bible), in recent years it also has become a visible and powerful component of broad public discussion. The published literature and numerous internet discussions show an exponential growth and intensification of this subject in recent years. Current political instability, nuclear proliferation, the growth of terrorism and intolerance, as well as demographic, economic and environmental uncertainties and tensions, are contributing to these gloomy predictions. If we add to this possible technological threats, like self-replicating nanoscale systems which may quickly consume 'ordinary' biological entities (the 'grey goo' scenario), or the possibility of an uncontrollable genetically engineered pandemic, the overall picture becomes very scary indeed.

In fact, in recent years the doomsday argument has evolved from a somewhat fringe and 'conspiracy theory' subject into a vocal issue of mainstream philosophical and futurological discourse. Such serious authors as, for example, the Canadian philosopher John Leslie (Leslie 1996) and the renowned British cosmologist and astrophysicist Martin Rees (Rees 2003) have summarized numerous arguments about why, more likely than not, our luck may soon be running out. Without going into a detailed review of all the various possible (competing and/or concurrent) scenarios for our possible extinction, it is sufficient to mention one, probably the most serious, threat. There is an unstoppable and ever-growing capability for small but dedicated groups of people (and even single individuals) to inflict large-scale damage, to which events like the Oklahoma bombing of 1995 and the horrors of 11 September 2001 bear ominous testimony. Due to unstoppable advances in nuclear engineering and technology, the sheer amount of physically destructive power potentially available to small clandestine groups is becoming increasingly frightening.

While some may downplay the above as conspiracy theories, the unfortunate reality is that the destruction of life on Earth through the explosion of a super-powerful nuclear device ('the doomsday machine') is not an unthinkable possibility. On the contrary, it is fully within the technological means existing today to construct such a device. Apart from nuclear threats, there are ongoing discussions about bio-terrorism, cyber-terrorism and other sinister means of mass destruction. Even such crazy scenarios as an attempt to deflect a large asteroid from hitting the Earth are not within the realm of the impossible (e.g. the recent blockbuster movies *Deep Impact* and *Armageddon* depict technologically plausible means of deflecting asteroids).

On top of all this, it is becoming increasingly clear that no functionally sensible social system can fully control all the fringe individuals who may be nourishing such megalomaniac suicidal plans for all humankind. If is not uncommon for single individuals to commit suicide (some available statistics suggest that, on average, as many as one per cent of all people die by suicide[2], and most likely not all suicide deaths are reported as such), why not take the rest of humankind with you[3]?

It is, of course, impossible to predict with full certainty whether the doomsday scenario (in whatever fashion) will indeed come to pass or whether humanity will find a safe route to avoid it and keep progressing indefinitely. The most optimistic

---

[2] See www.medicalnewstoday.com/articles/234219.php
[3] Note the recent case where a suicidal pilot deliberately crashed a passenger jet, killing not just himself but over a hundred others.

scenario foresees the future proliferation of humankind beyond this planet, perhaps even to galactic and cosmic levels, as some scientists are proposing (e.g. Dyson 1979, 1988). This does not necessarily limit future cosmic humanity to our present biological form; in fact, some authors propose a human–computer symbiosis ('cyber-humans'), or even a complete transfer of consciousness to robots (Kurzweil (1999), (2005), Moravec (1999)). Marvin Minsky, one of the pioneers of artificial intelligence, has expressed a thoroughly optimistic view: 'Yes, we will engineer replacement bodies and brains using nanotechnology. We will then live longer, possess greater wisdom and enjoy capabilities as yet unimagined' (Minsky 1994).

The ideas of transhumanism and posthumanism have their vocal enthusiasts as well as skeptics. These ideas are certainly unorthodox and controversial. Some critics fiercely oppose the very idea that human nature can or should be changed or tampered with in any way. The popular social philosopher and bestselling writer Francis Fukuyama went as far as to list transhumanism among the most dangerous ideas the world is presently facing. He dismissed transhumanism as: 'a strange liberation movement whose crusaders aim much higher than civil rights campaigners, feminists, or gay-rights activists [...] This movement wants nothing less than to liberate the human race from its biological constraints' (Fukuyama 2004).

It remains to be seen if humanity will be able to work out more balanced and constructive views on these issues or whether the clashes and divisions will continue, at least for the foreseeable future.

When I myself contemplate these issues (especially in the context of the 'isotopicity paradigm' discussed in this book), I tend to reconcile these two 'opposites' in myself: to see our world and our existence in it as 'real' reality (what a tautology) and a 'simulated' matrix Universe. Maybe, there is a some kind antinomy (unity of opposites) in these two ways of looking at the world and ourselves. Everything rests on polarities: day and night, life and death, love and hate, positive and negative—everything: from our feelings and emotions to positive and negative electric charges. And the quantum uncertainly principle speaks of a similar duality: position and momentum, energy and time, particles and waves, discrete and continuous, order and chaos. And the duality of information (both 'ideal' and 'physical') comes into play here as well.

From ancient times until the present day there has been an interesting (and often uneven) relationship between the 'twin sisters': 'real' physics and 'ideal' mathematics. And we most certainly need both of them as they need each other. Thus, perhaps, in some synthetic modality, we are both 'real' and 'simulated' in the overall infinite Platonic world of numbers and forms. And this is especially the case, if our contemplations in this regard move along the lines of such ideas as digital information, algorithms, binary strings (provided by isotopes and anything else), or the universal 'Library of Babel' of all possible books and all possible knowledge.

# References

Anderson P W 1990 On the nature of physical laws *Phys. Today* **43** 9
Anderson P W 1991 A question of mind over measurement *Phys. Today* **44** 146

Arnold A 1992 *The Corrupted Sciences. Challenging the Myths of Modern Science* (London: Paladin)

Berezin A A 1989 Isotopic engineering (perspectives) *J. Phys. Chem. Solids* **50** 5–8

Berezin A A 1996 Mainstream and fringe scientific ideas and ultimate values *Ultimate Real. Mean.* **19** 40–9

Berezin A A 2001 Discouragement of innovation by overcompetitive research funding *Interdiscip. Sci. Rev.* **26** 97–102

Berezin A A 2002 Energy, information, and emergence in the context of ultimate reality and meaning *Ultimate Real. Meaning* **25** 256–73

Berezin A A 2004 Ideas of multidimensional time, parallel universes and eternity in physics and metaphysics *Ultimate Real. Meaning* **27** 288–314

Berezin A A 2006 Simulation argument in the context of ultimate reality and meaning *Ultimate Real. Meaning* **29** 244–61

Bohm D 1980 *Wholeness and the Implicate Order* (London: Routledge)

Bostrom N 2003 Are you living in a computer simulation? *Phil. Q.* **53**(211) 243–55

Chalmers D 2005 The matrix as metaphysics ed C Grau *Philosophers Explore 'The Matrix'* (Oxford: Oxford University Press) pp 132–76

Cole S 1992 *Making Science: Between Nature and Society* (Cambridge, MA: Harvard University Press)

Condorcet C M J A N 1795, 1955 *Sketch for a Historical Picture of the Progress of the Human Mind* (New York: Noonday) (Engl. transl.)

Crawford I 2000 Where are they? Maybe we are alone in the Galaxy after all? *Scientific American* (July 2000) **283** pp 38–43

Dauben J W 1977 Georg Cantor and Pope Leo XIII: mathematics, theology, and the infinite *J. Hist. Ideas* **38** 85–108

Dauben J W 1979 *Georg Cantor: His Mathematics and Philosophy of Infinite* (Princeton, NJ: Princeton University Press)

Davies P 1993 *The Mind of God: The Scientific Basis for a Rational World* (New York: Touchstone)

de Bono E 1985 *Six Thinking Hats* (New York: Little Brown)

Dickson L 1952 *History of the Theory of Numbers* (New York: Chelsea)

Dickson L E 1960 *Modern Elementary Theory of Numbers* (Chicago, IL: University of Chicago Press)

Dyson F J 1979 Time without end: the physics and biology in an open universe *Rev. Mod. Phys.* **51** 447–60

Dyson F J 1988 *Infinite in All Directions* (New York: Harper and Row)

Egan G 1994 *Permutation City* (London: Orion)

Egan G 1997 *Diaspora* (London: Orion)

Feodorov N 1913 *Filosofiya Obschego Dela (The Philosophy of Common Deed)* (Moscow)

Feyerabend P 1975 How to defend society against science *Radical Phil.* **11** 3–8

Foltz F A 2000 The ups and downs of peer review: making funding choices for science *Bull. Sci. Technol. Soc.* **20** 427–40

Forsdyke D R 1993 On giraffes and peer review *FASEB J.* **7** 619–21

Forsdyke D R 2000 *Tomorrow's Cures Today? How to Reform the Health Research System* (Amsterdam: Harwood Academic)

Fukuyama F 2004 Transhumanism: the world's most dangerous idea *Foreign Policy* (September/October 2004) pp 42–43

Good I G 1969 Gödel's theorem is a red herring *Br. J. Phil. Sci.* **19** 357–8

Gordon R 1993 Grant agencies versus the search for truth *Accountability Res.* **2** 297–301

Haller E E 1995 Isotopically engineered semiconductors *Appl. Phys. Rev.* **77** 2857–78

Horrobin D F 1990 The philosophical basis of peer review and the suppression of innovation *J. Am. Med. Assoc.* **263** 1438–41

Horrobin D F 1996 Peer review of grant applications: a harbinger for mediocrity in clinical research? *Lancet* **348** 1293–5

Jahn R G and Dunne B J 1988 *Margins of Reality: The Role of Consciousness in the Physical World* (New York: Harcourt Brace Jovanovich)

Kenward M 1984 Peer review and the axe murderers *New Scientist* (31 May 1984 p 13)

King-Hele D G 1975 Truth and heresy over Earth and sky *Observatory* **95** 1–12

Kostoff R N 1997 Four factors and one criterion are key to improving peer review *Phys. Today* **50** 102–4

Kurzweil R 1999 *The Age of Spiritual Machines: When Computers Exceed Human Intelligence* (New York: Viking)

Kurzweil R 2005 *The Singularity is Near: When Humans Transcend Biology* (New York: Viking)

Lanza R and Berman R 2009 *Biocentrism: How Life and Consciousness Are the Keys to Understanding the True Nature of the Universe* (Dallas, TX: BenBella)

Lavine S 1994 *Understanding the Infinite* (Cambridge, MA: Harvard University Press)

Lederman L 1993 What can we learn from the supercollider's demise? *Scientist* **7** 12

Leslie J 1996 *The End of the World: The Ethics and Science of Human Extinction* (London: Routledge)

Lewis D 1986 *On the Plurality of Worlds* (Oxford: Blackwell)

Lloyd S 2006 *Programming the Universe: A Quantum Computer Scientist Takes on the Cosmos* (New York: Knopf)

Mann C C 1998 Who will own your next Good Idea? *Atlantic Monthly* (September 1998) pp 57–82

McTaggart L 2003 *The Field: The Quest for the Secret Force of the Universe* (HarperCollins)

Minsky M 1994 Will robots inherit the Earth? *Scientific American* **271** 109–13

Moravec H 1999 *Robot: Mere Machine to Transcendent Mind* (Oxford: Oxford University Press)

Oreskes N 1999 *The Rejection of Continental Drift* (Oxford: Oxford University Press)

Osmond D H 1983 Malice's wonderland: research funding and peer review *J. Neurobiol.* **14** 95–112

Penrose R 1994 *Shadows of the Mind* (Oxford: Oxford University Press)

Penrose R 1996 On gravity's role in quantum state reduction *Gen. Relativ. Gravit.* **28** 581–600

Pickover C A 1995 *Keys to Infinity* (New York: Wiley)

Pickover C A 2001 *Wonders of Numbers: Adventures in Mathematics, Mind, and Meaning* (New York: Oxford University Press)

Plekhanov V G 2004 *Applications of Isotopic Effect in Solids* (Berlin: Springer)

Plichta P 1997 *God's Secret Formula: Deciphering the Riddle of the Universe and the Prime Number Code* (Rockport, MA: Element)

Randall L 2005 *Warped Passages: Unraveling the Mysteries of the Universe's Hidden Dimensions* (New York: HarperCollins)

Rees M 2003 *Our Final Hour* (New York: Basic)

Rhinehart L 1971 *The Dice Man* (New York: William Morrow)

Ribenboim P 1989 *The Book of Prime Number Records* 2nd edn (New York: Springer)

Rousseau D L 1992 Case studies in pathological science *Am. Sci.* **80** 54–63

Rucker R 1987 *Mind Tools (The Five Levels of Mathematical Reality)* (Boston, MA: Houghton Mifflin)

Russell B 1989 *A History of Western Philosophy* (London: Unwin) (and many other editions)

Savan B 1988 *Science Under Siege: The Myth of Objectivity in Scientific Research* (Toronto: CBC)

Schmidt E and Cohen J 2013 *The New Digital Age: Reshaping the Future of People, Nations and Businesses* (New York: Knopf)

Sheldrake R 1988 *The Presence of the Past: Morphic Resonance and the Habits of Nature* (New York: Times Books)

Sheldrake R 2012 *Science Set Free* (New York: Deepak Chopra)

Shulman S 2000 Towards sharing the genome *Technol. Rev.* **103** 60–7

Szent-Gyorgyi A 1972 Dionysians and Appollonians *Science* **176** 966

Tegmark M 2014 *Our Mathematical Universe: My Quest for the Ultimate Nature of Reality* (New York: Knopf)

Teresi D 2002 *Lost Discoveries: The Ancient Roots of Modern Science—from the Babylonians to the Maya* (New York: Simon and Schuster)

Tipler F J 1989 The omega point as eschaton: answers to Pannenberg's questions for scientists *Zygon* **24** 217–53

Tipler F J 1994 *The Physics of Immortality: Modern Cosmology, God and Resurrection of the Dead* (New York: Doubleday)

Vinge V 1993 The coming technological singularity: how to survive in the post-human era *Vision-21: Interdisciplinary Science and Engineering in the Era of Cyberspace* ed G A Landis *NASA publication* CP-10129 pp 115–126

Webb S 2002 *Where is Everybody? Fifty Solutions to the Fermi Paradox and the Problem of Extraterrestrial Life* (New York: Copernicus)

**IOP** Publishing

# Digital Informatics and Isotopic Biology
Self-organization and isotopically diverse systems in physics, biology and technology
**Alexander Berezin**

# Chapter 7

## Conclusion. Message to the young reader

*Do not fear to be eccentric in opinion, for every opinion now accepted was once eccentric.*

Bertrand Russell (1872–1970)

The title of this conclusion as a 'message to the young reader' right away needs some qualification. Yes, I have a few words below that are addressed specifically to young(er) scientists and explorers who may still be in search for their big quest: what to do with their lives and the research lines they may be undertaking. And the choices and possibilities in this regard in our modern world are enormous, both in quantity and quality. But the 'young reader' in my view can also include persons (whether they are scientists or not) of any age who retain their passionate curiosity about the world and are, so to speak, 'young at heart and in mind' as, I humbly hope, the author of these lines is.

And I can say here (hoping that I will not be accused of vanity or bragging—or, even if I am, I can stand it) is that being now in my early 70s (I was born in 1944, one year before the end of WW II), my curiosity and questions about the world and our place in it have steadily increased (and not decreased) from my early days through all my path in science and they are still on the up-swing as of today. And I am more than sure that there are many (yes, thousands upon thousands) of people in my age group, and even older, who can state exactly the same about themselves. So, 'young people of all ages, unite'.

So, what is the prime message of this book? Let us recall how Richard Feynman was able to capture the most important cumulative result of science in one sentence (quoted in the first paragraph of the introduction of this book). The essence was that all the world is made up of atoms and all the rest can be unfolded from this primary fact. Likewise, what kind of sentence could summarize the main message of this book in the most concise way? At the risk of being accused of imitating

Feynman, I suggest that the key idea of this book is that 'the diversity of stable isotopies (isotopicity) may act as an additional informational factor in physical, self-organizational and biological dynamics and manifestations'. This book attempts to unfold this primary idea along several directions, as discussed in the chapters above.

And, of course, the primary devil's advocate argument objecting the above statement will be: 'if everything is as this author claims, why has nobody else put similar ideas into circulation until now?' Actually, this is the question that I was asked a number of times when presenting these ideas at various seminars and conferences.

While my short answer to this could be the story about the Mobius strip (section 6.4 above), my more detailed answer would be to go over several key points about big science, peer review, social and economic factors, competition for funds and the other factors that I detailed in chapter 6. How convincing my argument is is up to the reader to decide. In any case, I humbly believe that my book offers several lines of contemplation for enthusiasts and skeptics alike (and all grades in-between).

Repeating a few words from section 5.15, *if Nature is smart enough to use the diversity of chemical elements for biology (almost all the elements from the periodic table have some biological functions, including microelements), then it may look somewhat odd that Nature would omit to use such a mighty additional informationally rich resource as the diversity of stable isotopes for the structuring and functions of biological systems at all levels of evolution and complexity.* The likely 'answer' to such a 'puzzle' is that, yes, Nature most likely uses it (isotopic diversity) but we have so far failed to detect this and have even (largely) failed to look at it even at the level of hypothesis, not to mention any targeted experimentation. One of the primary aims of this book is to draw the attention of the world's research community to this incipient research area of stable isotopicity and isotopic engineering—a direction that (with some luck) may turn out to be a newly found gold mine for physics, biology, biomedicine, material science, cognitive sciences and informational technology in a broader sense.

As the author of this book, I realize that some readers will consider the ideas and suggestions presented here to be mere fantasies and speculations that never can or will be implemented at the practical level. Yes, I am familiar with such attitudes from my numerous seminars and coffee conversations. However, I have actually taken them as (unintended, perhaps) compliments as they, paradoxically, put me in good company. When Jules Verne published *De la Terre à la Lune* (*From Earth to the Moon*) in 1865 who really believed that travel to the Moon one day would become a reality? Well, it did, 104 years later. And when in 1903 Konstantin Tsiolkovsky suggested that a hydrogen–oxygen rocket (rather than the huge cannon of Jules Verne) could do the job, he was much closer to the target, but he was still largely seen as a visionary and a dreamer. Yet these people (H G Wells and a few others can be added here) planted seeds in the public perception that eventually stimulated technological and engineering developments towards successful implementations of these ideas.

Turning to the potential of isotopic informatics in physics, engineering, nano-technology, or biology and biomedicine (as described at length in this book), we can state that, yes, at the present stage we do not yet have all the technology needed for the practical testing of the ideas of stable isotopicity. Yes, we have technology for isotope separation, but to handle isotopes atom by atom (as some of the potential applications treated in this book call for) we need further advancement of such tools as atomic force microscopy, molecular beam epitaxy, micro- and nano-laser instrumentation and perhaps some other methods that this author is presently unaware of (but some others may well be). If this book stimulates further thinking and contemplation about all these isotopic opportunities, the author will see it as an indication that his effort brought some worthy result.

However, there are a few words that I would like to address specifically to young people who may still be university or college students (be it at the undergraduate or graduate levels), or who may be postdocs and research fellows at academic or industrial facilities, or working at start-up business enterprises, or perhaps in some other niche. Since I myself was in most of the above categories, I hope I can reliably attest to the typical aspirations, ambitions and frustrations that usually come along at these stages of life and career. Yes, many of you have dreams and desires to really make a difference in this world, but you are not in a pure vacuum but always within some structure (academic or whatnot) that brings some constraints and rules of conduct with it. If you are actively involved in a research project, you are most likely a part of some team (research group, industrial r&d lab, or whatnot) and you have some supervisors and/or superiors ('bosses') with ongoing research programs and you have to find your 'freedom within this matrix'. While I certainly cannot give an exhaustive 'one size fits all' algorithm on how best to navigate in such an environment, I would still like to give you some tips based largely on my 50+ years of working in science. Take them not as the preaching of some guru sitting somewhere on a mountain top, but simply as helpful hints from someone who traveled similar routes before you.

(1) First of all, form as clearly as you can, your own vision of your interests and goals. Put them in writing in a special notebook. You do not need to do this all at once, but be sure that you keep all the records in a well-organized and dated way. You can have several directions and levels for your interests and it is okay to add to the list of your interests and goals and revise your priorities. (And one more comment from me: in spite of the fact that nowadays most of our recording and writing goes on electronically on laptops, USB sticks, etc, I strongly recommend for practical and emotional reasons that you keep the habit of paper notebooks for your most important ideas and plans. Printouts are okay too, if you keep them organized and stored in binders.)

(2) If you have some specific idea or hypothesis, no matter how far-fetched and speculative it may appear to you, try to shape it up in the form of a short draft of an article—the chances are that it may indeed be a seed for publishable material. If you have some trustworthy friends and/or colleagues who may be interested in your idea, it is all right to collaborate

(and perhaps later to co-author), but be sure that your idea is in no way compromised by such collaboration. This is a fine art, of course, and not all scenarios can be predicted here, but it is important that you maintain full control over the spelling out of your ideas.

(3) If you are a part of a research group (e.g. if you are a graduate student or postdoc) and your idea 'does not fit' the research program you are in (or, to put it flatly, your supervisor ('boss') does not share or support your idea), then you are in a situation that I was in more than once during my career (as, I am sure, many others have been too). If this is currently what you have, you should figure out what the best way for you to navigate in such a situation is. If your idea has already shaped up to the level of a research paper, you may consider publishing it over the head of your supervisor. In the university environment and in groups that are not involved in any classified or proprietary projects, such a practice is rather common—at least, I did it a number of times and never regretted doing this. A good share of my earlier papers (which are included in the references of this book) were like that—concerning work that I did and published on my own initiative and that was outside of the formal work that I was paid to do. Of course, I did the latter work too, otherwise I would not have received my paycheck (or rather, cash—there were no checks in the USSR at that time and all wages and transactions were in paper cash). Remember that your ideas are far more important than being a 'good boy/girl' in the eyes of this-or-that boss. And if you have to choose, always be sure that your choice comes from your heart. Fortunately, the above situation becomes much easier for people who obtain a teaching position at a college or university. In this case, there are virtually no restrictions on what people can do for their research, especially if it is theoretical or conceptual work. If the teaching for which you are paid goes well, the university is unlikely to care too much what you research and publish (although, like any bureaucracy, they practice 'paper counting' for promotional and other administrative reasons).

(4) Do not be upset or put down by people who disagree with you and may be voicing their disagreement to you, sometimes quite aggressively. Such reactions are typical and quite common. As Robert Kennedy put it 'One fifth of all people always disagree with anything'. Whether we like it or not, this urge for confrontation and putting others down (those who dare to have their own views and—God forbid—ideas) is pretty much a common trait of our biological species. This does not mean that you should see yourself as a saint, immune to the above (none of us can safely make such a claim), but the best you can do is to take it calmly and politely, on a 'let us agree to disagree' note. And even if you are angered by what your opponents say, never show any aggressiveness or scorn in return.

(5) Also look carefully to see if there may be some real substance in your opponents' arguments and criticism. If so, consider their comments in a creative and constructive way. That may happen, not very often in my own experience, but occasionally it does. And as for Kennedy's 'one fifth', my

own experience in academia and in dealing with peer review many dozens of times makes me want to upgrade his estimate to a far higher level, perhaps 4/5 at least. But never mind, if you are persistent, consistent and stubborn (it is all right to be stubborn about good ideas), you will make it through. Remember what Winston Churchill used to repeat: 'Never, never, never give up!'

(6) Do not be scared by the peer review system. Read more about it in chapter 6 of this book and also look at what numerous posters on the web have said about it. You will find a lot of criticism of it, as well as some ideas about how to get around it. Yes, it is possible to 'cheat' the peer review system (as many authors have before) and get your most important ideas published. In fact, in the electronic age we are currently in (web pages, blogs, YouTube, etc—and the list is growing) there are many alternative ways to circulate your ideas outside the formal system of science publications. You can even start your own online journal or self-publish a book. The latter option (self-publishing) needs to be studied carefully in each particular case, because this industry contains numerous predators and scams, yet many people go along this route with various degrees of success.

(7) And finally, have your own list of role models—people whose lives and work encourages and uplifts you. People who came through struggles, frustrations and misunderstandings before they made their difference in history. People like Nikola Tesla, Marie Curie, Alfred Wegener (continental drift), Georg Cantor (the theory of infinite sets), or such great martyrs of science and ideas as Giordano Bruno or Hypatia (an ancient female mathematician, astronomer and philosopher, who was brutally killed by a mob in Alexandria in 415). You do not necessarily need to expect to join this list, but the lives of these people (and you can find many more to make your own list) can inspire and offer insights for your own ideas and endeavors.

These are my short hints to you, the reader, and in travelling along your own path in science, you will most likely be able to add your own thoughts and ideas to the above reflections. Good luck.

Lightning Source UK Ltd.
Milton Keynes UK
UKHW05n2321210318
319819UK00004BA/124/P